Gx 3/10/06 18
Lateover Ed.
ULlevel/

AS/A-Level Geography

Coursework & Practical Techniques

David Redfern

Malcolm Skinner

Philip Allan Updates
Market Place
Deddington
Oxfordshire
OX15 0SE

tel: 01869 338652
fax: 01869 337590
e-mail: sales@philipallan.co.uk
www.philipallan.co.uk

© Philip Allan Updates 2002

ISBN 0 86003 750 9

Printed by Raithby, Lawrence & Co. Ltd, Leicester

Contents

Introduction iv

Your investigation 1

What is a geographical investigation? 2

How do the examination boards assess investigative skills? 3

Establishing a suitable title for your investigation 5

Collecting primary data 8

Collecting secondary data 16

Presenting your results 18

The analysis and interpretation of data 30

Writing up the report/coursework 40

Written alternative examinations 47

Fieldwork-based/practical — AQA (A) Unit 7 49

Fieldwork-based questions 57

Practical questions 62

Introduction

The requirement to undertake geographical coursework at A-level has been present for many years, but this does not mean that students find the organisation of such work easy to cope with. The AS/A2 specifications in geography all state that fieldwork and the subsequent completion of a geographical investigation are necessary for all students.

The assessment of this investigation takes one of two routes:
▶ a piece of written/word-processed work that is submitted for assessment to either a tutor, a moderator or an examination board
▶ a series of activities that are assessed in a written examination

The purpose of this book is to guide you, the student, through both of these alternatives.

The first section deals with the nature and process of an investigation. It explains how to establish a suitable title, collect data, present and analyse material and how to write up the final report. Some specifications also allow a student to write about a fieldwork investigation he/she has completed in an alternative written examination, without the formal submission of the completed piece of work. This section will help with this process too.

The second section looks in more detail at the various types of written alternative examination that the examination boards offer. Here, students are not only required to write about a piece of fieldwork they have completed but also to demonstrate their ability to use a wide range of investigative skills in an examination context. In this section, example questions are provided, together with sample answers and examiner's comments.

You may choose to read this book from cover to cover or, more likely, you will read the relevant section when you come to that point in your course. It is intended to act as a personal guide or tutor, enabling you to see where to go next and, ultimately, how to maximise your marks. We hope that this will keep you on the right track for success.

AS/A-Level Geography

Coursework
& Practical
Techniques

Your investigation

Your investigation

What is a geographical investigation?

All AS/A-level geography specifications require you to undertake investigative work based on evidence from primary sources (including fieldwork) and secondary sources. In most cases, this work will be based on a combination of both primary and secondary source material.

In simple terms, primary data are those collected by you in the field or material from other sources which needs to be processed. Secondary data are derived from published documentary sources, and have been processed.

Primary research is a must — you must have had some direct contact with the area of study and/or the subject under investigation. This could involve a specific economic activity, an identified group of people or a local issue. You need to have visited an area and talked to the people or recorded other data there.

All geographical investigations should follow the same stages of enquiry:
▶ the identification of the aim of the enquiry, often in the form of testing a hypothesis or establishing research questions
▶ the collection of data, for example by measuring, mapping, observations, questionnaires or interviews
▶ the organisation and presentation of the data using cartographic, graphic or tabular methods, with the possible use of ICT
▶ the analysis and evaluation of the data, noting any limitations
▶ the drawing together of the findings of the investigation, and formulating a conclusion
▶ the suggestion of further extension to the investigation, including additional research questions that may have been stimulated by the findings
▶ a statement of success or otherwise of the investigation, with some commentary on the significance of the investigation for others

The geographical investigation therefore provides an opportunity for you to demonstrate what you can do outside the examination room.

It also allows you to build up evidence for the achievement of all of the Key Skills, and enables you to submit a portfolio for the Key Skills qualification. Your investigation could demonstrate aspects of the following Key Skills:

▶ **Communication**
  - read and synthesise information
  - write different types of document
▶ **Application of Number**
  - plan and interpret information from different sources
  - carry out multi-stage calculations
  - present findings, explain results and justify choice of methods
▶ **Information Technology**
  - plan and use different sources to search for and select information
  - explore, develop and exchange information, and thus derive new information
  - present information including text, numbers and images

How do the examination boards assess investigative skills?

There are a variety of ways in which the examination boards assess investigative skills. These are listed below.

▶ A piece of coursework that entails you completing an assignment/project/report. This will be a lengthy document (the length of which will be stated in the specification you are studying), usually presented in a lightweight folder.
▶ A written examination paper which seeks to test both fieldwork skills and other investigative skills.
▶ A written examination paper, based on pre-release material, which seeks to test investigative skills based on both primary and secondary data.

Most of the examination boards make use of more than one of these ways in their assessment programme, and it is essential that you check the requirements of your particular board. Table 1 provides a summary of the requirements of the six Geography specifications in England.

Check your specification for the following:
▶ In which assessment units are investigative skills assessed?
▶ How long should the coursework be, and will there be a penalty for it being too long?

Table 1 Different requirements for the coursework/skills units for each of the major specifications

	AQA Specification A	AQA Specification B	OCR Specification A	OCR Specification B	Edexcel Specification A	Edexcel Specification B
Which units and when?	Unit 3 (written) — AS; Unit 6 (coursework) or Unit 7 (written) — both A2	Unit 6 (written) or Unit 7 (coursework) — both A2	Unit 2682 (written) — AS; Unit 2685 (coursework) or Unit 2686 (written) — both A2	Unit 2689 (written) — AS; Unit 2690 (coursework) — A2	Unit 3 (coursework) — AS; Unit 6 (written) — A2	Unit 3 (coursework) — AS; Unit 5 (part B) (coursework) — A2
Length of coursework	4000 words	3500–4000 words	Unit 2682: 1000 words; Unit 2685: 2500 words; Unit 2686: 1000 words	Unit 2689: 1000 words; Unit 2690: 2500 words	2500 words	Unit 3: 2500 words; Unit 5 (part B): 1500 words
Restrictions on coursework topic	No restrictions on topic area. Must contain primary data collection and be based on small area of study	Must link to AS or A2 content. Must be an enquiry into a geographical issue. Must contain primary data collection	Unit 2682/6 — no restrictions on topic area, but at local scale. Unit 2685 — normally related to some area of specification	Unit 2689 — content chosen from 2687/2688. Unit 2690 — either based on fieldwork data, or significant use of ICT for the gathering of data	Must link to AS or A2 content. Must contain primary data collection	Unit 3: small-scale area linked to an environment studied for AS Individual Research Action Plan required. Unit 5 — title selected by student, and presented as a formal essay. Reports should be word-processed
Assessment of coursework	Marked by awarding body. Advisory service for teachers	Marked by centre, moderated by awarding body. Advisory service for teachers	Marked by awarding body	2689 — marked by awarding body. 2690 — Marked by centre, moderated by awarding body	Marked by centre, moderated by awarding body. Prior approval required	Unit 3 — marked by centre, moderated by awarding body. Unit 5 — marked by awarding body. Prior approval required for both
Written units	Unit 3: Two themes, either physical or human, plus questions based on fieldwork. Unit 7: Topic area announced 2 years in advance	Unit 1: Contains some fieldwork-related questions. Unit 6: Two compulsory questions, one based on a fieldwork enquiry, one based on skills	Unit 2682: Skills-related questions, with some linked to 1000-word report submitted at time of examination. Unit 2686: Skills-related questions, with one linked to 1000-word report submitted at time of examination	Unit 2689: Skills-related questions, with one linked to 1000-word report submitted at time of examination	Unit 6: Evidence of fieldwork should be included in answer to one question	Not applicable
Pre-release material	Unit 7: 4 weeks in advance of examination	Unit 5: the synoptic unit 4 weeks in advance of examination	Not applicable	Unit 2692: sustainable development	Not applicable	Unit 6: the synoptic assessment 2 weeks in advance

▶ Is there a specified format in which the coursework should be submitted?
▶ What are the restrictions on the titles or areas of study?
▶ How will the coursework be marked?
▶ Which specifications and assessment units offer a written alternative to the coursework?
▶ Which assessment units are based on pre-release material?

The proportion of the final marks that the investigative work carries varies from 15 to 20%, depending on the specification. Whatever the proportion, this section of marks is within your control. The purpose of this book is to guide you through the investigative process, whether in the form of coursework or a written alternative. Take advantage of this opportunity to boost your performance, and maximise your overall grade.

Establishing a suitable title for your investigation

Hypotheses, research questions, aims and objectives

The best type of investigation involves the testing of a **hypothesis**, or the setting of one or more **research questions** that can be investigated and evaluated. The hypothesis should be a statement based on a geographical question. For example, the question 'Do the characteristics of a soil change down a slope?' could form the basis for any of the following hypotheses:
▶ the depth of a soil varies downslope
▶ the clarity of the soil horizons varies down the slope
▶ the texture of a soil varies according to the position on the slope
▶ the acidity of a soil varies downslope
and so on.

The statement, or hypothesis, can now be tested by the collection of data that are both easy to identify and to collect. Once investigated and evaluated, the results of the enquiry can then be used to answer the original question, with reference to that aspect of soils being investigated.

The following examples illustrate the differences between hypotheses and research questions:
▶ A poor project title would be: *Changes in the retail geography of a town over the last 30 years.* Here there is no question, or hypothesis.
▶ A better title would be: *What have been the major changes in shopping*

patterns in the area around Wakefield within the last 30 years? This establishes a question to which an answer can be developed.

▶ An alternative title, in the form of a hypothesis, would be: *The CBD of Wakefield has seen major decline in activity within the last 30 years, caused by the growth of out-of-town retail outlets.*

Either of these last two approaches is an acceptable way in which to proceed.

The purpose of these approaches is to enable you to clarify your **aims** and **objectives**.

Aims are statements of what you hope to achieve. In the above case, you will want to identify areas of change — areas of decline of some shopping areas, and growth of others — with possibly some reasons for these changes.

Objectives are statements of how you will achieve your aims. What data will you need in order to identify areas of decline/growth? How can these data be collected, analysed and presented? How will you obtain evidence of reasons/causes? Do you need to write to anyone, interview anyone or devise a questionnaire? What precise form of fieldwork needs to be undertaken?

What is a null hypothesis?

Some investigations are best suited to the approach of establishing a **null hypothesis**. This takes the form of a negative assertion which states that there is *no* relationship between two chosen sets of variables. For example, a null hypothesis could state that *there is no relationship between air temperatures and distance from a city centre*. An alternative hypothesis can also then be established, namely that *temperatures decrease with distance from a city centre.*

The null hypothesis assumes that there is a high probability that any observed links between the two sets of data (temperature and distance from a city centre) are due to unpredictable factors. If temperatures are seen to decrease with distance from the city centre in the chosen city, then it is a result of chance. However, if the null hypothesis can be rejected statistically, then we can infer that the alternative hypothesis is acceptable.

One benefit of this seemingly reverse approach to an enquiry is that if the null hypothesis cannot be rejected, then it does not mean that a relationship does not exist — it may simply mean that not enough data have been collected to reject it. In short, the investigation was not worthless, but was just too limited in scope. Another benefit of this approach is that it allows the use of statistical tests on the significance of the results to be carried out.

Finally, it is accepted that this is a sophisticated way in which to approach an enquiry, and should only be used when fully understood.

Checking the feasibility of your investigation

Before embarking on this major piece of work which will, if successfully done, take a great deal of your time to complete, you should check whether or not the tasks involved are achievable, and that an overall conclusion will be forthcoming.

Here are some questions you should ask yourself to see if your idea can really work.

Is my topic area within the requirements of the specification?

You need either to check the specification yourself or to ask your teacher, or you may have to seek prior approval from a coursework moderator appointed by the examination board.

Is the subject matter narrow enough?

In general, it is better to study one aspect in detail than several aspects sketchily – to study one town rather than three. Many investigations are best suited to a local area of study – one that is large enough to give meaningful results, but not too large to become unmanageable. Can you visit the area of study (you may have to on more than one occasion)? Again, seek guidance from your teacher on the scale of study for your chosen topic.

Will I be able to collect the data that I need?

Remember, the use of primary data is paramount, and a whole range of sources of primary data exists – both quantitative and qualitative (see below).

Quantitative	Qualitative
Land use transects	Questionnaires
Housing surveys	Interviews
Environmental impact assessments	Field sketches
Traffic counts	Photographs (taken by you)
Climate surveys	
River measurements	
Soil surveys	

As the examination boards place great emphasis on data collection, it is wise to use a variety of data sources for your investigation. Do not just use questionnaires.

Will I be able to complete the investigation within the time period allowed?

One type of investigation that needs to be carefully monitored is that which deals with changes over time. In other words, one that needs a before and after element. Examples of this would include the impact of a new bypass around a settlement, or the impact of a new retail park. Check whether or not data within the two time periods can be collected.

What equipment will I need?

The amount and nature of the equipment will obviously depend on the nature of the data to be collected. However, you do need to check whether the equipment is available and in working order.

What else do I need to do?

Here is a final check list of some of the main requirements of any fieldwork or data collection:

▶ **Always ask for permission if it is needed** — any investigation that involves you entering someone else's land or building will require permission from the owner. In most cases, a letter or telephone call explaining the purpose of your visit will suffice. However, it is often better to ask your teacher to write a short note in support of your work on school/college-headed notepaper.

▶ **Check that you can get to your area and that you can get back** — is there public transport to the area or will you need help? Note that much fieldwork is undertaken at weekends, and Sunday timetables are often different.

▶ **Wear appropriate clothing** — if you are going into remote and difficult areas, then you must wear clothing that is warm and dry. Take special note of footwear, even in urban areas.

▶ **Be safety conscious** — *never* work alone and always tell someone where you are working. If possible, give details of times of return and methods of travel. Be particularly careful in coastal areas. Check on tide levels, never work below crumbling cliffs and stay on coastal paths.

Collecting primary data

Once you have established the aims and hypotheses of your investigation, you need to think of the data that are required and what method of collection will be necessary. Investigations at this level are essentially based upon your own observations, which involve collection of primary data through such techniques as questionnaires, interviews, river measurements, pedestrian surveys and urban transects. The whole of your investigation can be based on

such material but, depending on the scope of your work, it should be possible to include some secondary data from previously published sources.

Sampling

Sampling is used when it is impossible, or simply not necessary, to collect large amounts of data. Collecting small amounts of carefully selected data enables you to obtain a representative view of the feature as a whole. You cannot, for example, interview all the shoppers in a market town or all the inhabitants of a village, but you can look at a fraction of those populations and from that evidence indicate how the whole is *likely* to behave.

When you have established the need for a sample survey, you then have to decide on the method that will ensure a large enough body of evidence is collected objectively. If, for example, you are interviewing the inhabitants of a village which clearly consists of people of a variety of ages, it is no use holding the bulk of your interviews with young members of the opposite sex as this will distort your picture of the settlement. There are a number of sampling techniques, the main ones to be considered here being random, systematic and stratified (quota).

Types of sampling

A **random sampling** is one that shows no bias and in which every member of the population has an equal chance of being selected. This method usually involves the use of random number tables.

In **systematic sampling**, the sample is collected in a consistent manner by the selection, for example, of every tenth person or house. On a beach, you could decide on sample sites every 100 m along the feature and in each location use a quadrat in which you select the pebbles at intersection points within the grid, at regular intervals.

Stratified sampling is based on knowing something in advance about the population or area in question. For example, if you are surveying a population and you know its age distribution, then your sample must reflect that age distribution. If an area is being surveyed in which you know the distribution of soil types, then sites should be taken in proportion to the area covered by each type of soil. For instance, if a particular soil covers 40% of the area, 40% of the total sample points should be taken within the area covered by that soil.

Bias in sampling

In sampling it is possible, through poor choice of method or insufficient evidence, to achieve a result that is unrepresentative of the population in question. Taking samples on the same day of the week or outside the same shop could lead to a distortion in a shopping survey, for example.

The size of the sample

The size of sample usually depends on the complexity of the survey used. With a questionnaire, it is usually thought necessary to interview sufficient people to take into account the considerable variety introduced by the range of questions used. Sample size can be restricted by practical difficulties which can act as a limit to collection and the reliability of results. Your aim should be to keep the sampling error as small as possible. You are not a professional sampler and cannot be expected to conduct hundreds of interviews, but on the other hand, sampling only 20–30 people in a market survey would just not be representative of the population as a whole.

Point sampling

Such sampling is carried out in surveys which involve, for example, studies of land use, vegetation cover, soil sites and selection of such items as pebbles in longshore drift studies. Point sampling involves the use of a grid — that produced by the Ordnance Survey is ideal, but for field surveys a quadrat can be used. A quadrat is a frame enclosing an area of known size (often $1\,m^2$), and is subdivided by the use of wire or string. Both random and systematic sampling can be carried on within this framework.

Questionnaires

Many investigations at this level involve the use of questionnaires, the writing of which can be one of the most time-consuming and difficult aspects of an individual investigation.

Questionnaires are often seen by examiners as badly designed, with an insufficient balance between the more specific questions and the more open-ended type. Such open questions can certainly produce material that is less quantifiable, but they can provide additional information that is vital in trying to explain behaviour as revealed in the answers to the more direct questions.

Questionnaire surveys, for example with supermarket/hypermarket studies, often involve a lot of effort but reveal little more than very basic information on the behaviour of shoppers. Oddities in the pattern may not be accounted for because there was a failure to include open questions which may have gone some way to understanding shoppers' motives for their behaviour.

Questionnaires can become far too sociological, and therefore of limited geographical value, by not concentrating on the spatial aspects of the sample being studied. Questionnaires that represent little more than a social survey should be avoided as they will give little scope for mapping and analysis of patterns.

At GCSE level, you were often led to undertake a questionnaire survey in order to demonstrate your ability to complete it. At this level, the aim is

completely different; quality, quantity and the reliability of the data produced are as important as the means by which they were obtained.

When attempting a questionnaire survey, there are a number of guidelines that can be followed:

▶ Keep it as simple as possible — busy people do not like answering a lot of questions.
▶ Try to write a mix of closed questions (yes/no answers or multiple response) and open questions (choice of answers or free statements).
▶ Decide on a sample size that is adequate.
▶ Always try to introduce your questionnaire in the same way (write a brief introduction).
▶ Put the questions in a logical sequence.
▶ Ask questions that will produce data that can be analysed.
▶ Think carefully about sensitive questions and use tick boxes for such information if possible.
▶ For respondents' ages and other such information it is often better to offer categories rather than insist upon an exact figure.
▶ Try to ask questions about a person's behaviour and not how they perceive their behaviour.
▶ Pilot your questionnaire by testing it out to see if it will produce the material that you want.
▶ Always seek approval from your teacher or tutor before proceeding, in order to avoid insensitive questions and also to prevent harassment of local people by swamping the area with too many questionnaires.
▶ Obtain a document from your school/college that states exactly what you are doing.
▶ Always be polite, look smart and smile, and do not get upset if people refuse to answer.
▶ *Never* work alone.
▶ If you intend to stand outside a specific service or in a shopping centre, it is a good idea to ask for permission.

Interviews

In the course of your work, it may be necessary to conduct an interview in order to see how some people stand on an issue or how they would act in certain circumstances. In an investigation of rural land use, for example, it might be a good idea to interview farmers, which may help to account for some of the changes that you have observed. If you are investigating a conflict as part of your work, then it is essential that you should try to obtain the views of the parties involved. This will certainly take you beyond the scope of a questionnaire, as you will need to put different questions to each respondent or group involved. It is not always possible to meet the people involved, so you might have to write to a person or group for their opinions.

Interviews give you the opportunity to explore more detailed areas and to follow up any points that are raised. They can be time-consuming, so you are advised to think carefully before deciding upon such a course of action, and if you decide it is a good idea, use the technique sparingly.

Always present a well-written letter asking for the interview, and prepare for it well, as people do not like wasting their time. Being prepared means that you should have a list of direct questions that need to be asked, making sure that the person to whom you are speaking has the information you require.

Do not ask questions where it is possible for you to obtain the information for yourself. Take notes at the interview or immediately afterwards. You could record the interview, but always obtain permission before you start. Your final presentation can be very much enhanced by quotations from interviews, but always check accuracy and ask permission from the interviewee.

Sampling attitudes

Collecting information on peoples' attitudes is often quite difficult. You can ask about the attitude of an individual during an interview, but if your course of action is to obtain a number of responses through a questionnaire survey, you must phrase questions in such a way as to avoid bias creeping into your final results. There are three main ways of collecting such information.

▶ **Bi-polar tests** involve establishing a rating scale based on two extremes of attitude (i.e. poles apart). For example, on a CBD survey you could ask shoppers if they found the shopping area attractive or ugly and offer them a sliding scale of response from (1) Ugly to (7) Attractive. You could put a number of such points to them and as a result give the area an overall rating. This would be useful when comparing shopping centres.

▶ A **point-score scale** is used to give respondents the chance to identify factors that they consider to be important. You could ask respondents to place each factor on a scale from 0 to 4 in terms of importance, for example in attracting shoppers to a retailing centre.

▶ Using a **rating scale** allows respondents to agree or disagree with a statement. Statements could be put to them and then they could be asked to place their response in categories ranging from 'Strongly agree' to 'Strongly disagree'.

Types of survey

So far we have mainly shown you how to collect information by asking people questions. There are other types of survey that do not involve people

and that you will find useful in obtaining information. Here are some of them, with details, in many cases, of how to proceed.

Surveys in urban areas

A large and varied number of surveys can be carried out within the context of urban fieldwork. Some of these are outlined below.

▶ **Land use surveys** can be used for both rural and urban areas and must always be carried out with a clear purpose in mind. Land use should be placed into categories where groups of similar land uses can be identified. Sub-groups within categories should be established depending upon the detail required. With housing, for example, several categories can be recognised, such as terraces, detached, bungalows, semi-detached and flats.

▶ **Land use transects** are used when the urban area is too large to survey as a whole, so a sample of it has to be taken. The transect is essentially a slice through the urban area to see how land use varies from one part to another. The usual transect will start in the middle of the urban area and run along a radial road to the urban fringe. Such surveys are often used to show how the land use on one side of an urban area differs from another. It can also be used to show where CBD functions cease and others take over, and is therefore used to delineate that central area. As well as land use, other information collected on a transect could include building height, number of storeys, upper floor use, building condition (using an index of decay) and the age of buildings. It is also possible to collect information not concerned with urban geography in this way. Such projects could include noise and pollution surveys, and temperature/humidity readings across an urban area.

▶ **Environmental surveys** can be carried out as some form of appraisal or assessment where the use of a point-score scale is recommended. In such cases, both positive and negative observations can be made. Environmental surveys can be carried out in urban areas, on beaches and within river channels and valleys. This can be done with first-hand observations and by interviews with local residents or, in the case of recreation and tourist areas, visitors. Noise, water and air pollution can also be studied. It is possible to use instruments and chemical testing kits in such surveys, but there are simple tests that you can undertake using your own observations. Noise pollution can be estimated using a simple table, as long as the same person makes all of the observations. Air pollution can be assessed with the aid of a roll of sticky tape with which you can take samples from different surfaces in various parts of a town. Water pollution can be assessed visually (or by smell!), but it is possible to use a chemical kit or a secchi disc. The latter is a disc that can be lowered into the water and the point recorded at which you are no longer able to see it and at which it comes into view again when lifted.

► **Shop location/shopping quality surveys** can be carried out in the CBD or even in the urban area as a whole. Shop location studies usually take place in the central area with the aim of finding distinct patterns of land use within the CBD. Selected shop categories can then be analysed using the technique of nearest neighbour analysis which indicates the degree to which the category is clustered. You could calculate an index of dispersion. (Such surveys can also be carried out on service outlets and offices within the urban area.) Shopping quality surveys can involve observational data or the use of a questionnaire.

► **Land value/house price surveys** are often carried out to identify the peak land value point (PLVP)/peak value intersection (PVI) in an urban area. Taking the values of properties from estate agents and newspapers is much easier than trying to find the values used for local tax assessment (these can be obtained from the local authority's valuation office where these records are open to inspection). Values should be converted into a unit per square metre of ground floor space.

► **Traffic flow surveys** can be carried out by measuring the traffic flow past several survey points within the urban area.

► **Pedestrian flow surveys** are one of the recognised ways of indicating commercial activity within a CBD. There are a number of points to remember when contemplating such a survey:
  - you will not be able to carry this out by yourself
  - mornings and afternoons are best, so avoiding the movement of office workers that takes place in the middle of the day
  - do more than one survey in order to contrast different times of the day or even of the week
  - at busy points, use two counters operating, if possible, back-to-back in the middle of the pedestrian throughway
  - shopping centres are usually private property, so it is advisable to ask for permission

Surveys in physical geography

Within the realm of physical geography there are a number of surveys that can be carried out. Some of the more popular studies are outlined below.

► **River surveys** involve the taking of measurements that are then used in the calculation of fluvial features such as discharge, load, friction and efficiency. The most popular seems to be the calculation of the discharge of a river which involves finding the cross-sectional area at certain points and multiplying it by the calculated speed (usually obtained from a flow meter) to give a figure expressed in cumecs (cubic metres per second). The calculation of the cross-section also identifies the wetted perimeter. Other calculations or measurements that can be made include the gradient of the stream and the shape and size of the bed load. A final

measurement to consider is the extent to which the river meanders — its index of sinuosity.

▶ **Surveys of slopes** usually involve calculating the steepness by means of a clinometer, measuring tape and ranging poles. One of the features of slopes that it is possible to measure is the amount and type of vegetation present, information that can be displayed on a kite diagram.

▶ **Soil surveys** can be carried out on slopes, choosing sites in exactly the same way as you would for slope vegetation surveys. Various tests can be done, including: soil acidity (using a chemical soil testing kit); soil texture (using a technique of feeling the texture with your fingers to find the sand, silt and clay elements); and moisture and organic content (by taking a sample, weighing it, drying it overnight and weighing it again, burning off the organic matter to leave the inorganic matter, the weight of which can be compared with that of the original sample).

▶ **Coastal surveys** could involve measuring the direction and amount of longshore drift and examining the structure of sand dunes, including the environment for plants and animals.

▶ **Glaciation surveys** in the field usually involve a study of glacial deposits called till fabric analysis. This is based on the idea that stones within the ice will become orientated in a direction that presents minimal resistance. It means that they should be found in glacial deposits with their long axis parallel to the direction of ice movement. From this you should be able to establish the direction of ice movement and possibly its source.

▶ **Microclimate surveys** can be carried out in your local area possibly from rural to urban areas or across an urban area. Other contrasts could be made between day and night conditions and, more ambitiously, between seasons.

Field sketches and photographs

Both of these methods are excellent ways of recording observations that you have made as part of your investigation. Field sketches enable you to pick out those features within the landscape that you consider important. Investigations in physical geography lend themselves nicely to this technique, particularly those on coastlines and glaciation. It does not matter if you cannot draw particularly well; it is far more important to produce a clear drawing with useful annotations.

If you are not that confident in your drawing ability, then try photography as an alternative. Remember, however, that far too many photographs find their way into the final presentation as attractive space fillers in the mistaken belief that they will make it 'look good'. You need to select carefully the images that you want to show. As with sketches, annotation is vital, as photographs are only useful if you point out to the reader the major features that you have observed. Photographs should be included at the relevant points in the text and not in a large block, which sometimes makes it very difficult for the reader to see their purpose.

Collecting secondary data

Secondary data collection involves the gathering of data that have already been put into written, statistical or mapped form. For an investigation at this level, there is a wide range of sources that can be accessed.

If you are involved in an investigation that has a temporal context, it is almost certain that you will need to access secondary data from previous surveys in the course of your work. Secondary material can be very useful in the early stages of an investigation, where it can provide a helpful context. It can be used too when explaining and discussing primary material. You might also find it useful to combine field data with those obtained from newspapers, maps, census returns, local authority and other secondary sources. This should give you a much wider database for your analysis and for comparing first-hand material recorded in the field with other results previously obtained.

When using secondary data, it is important to check the accuracy of the material, particularly if the information could be biased. Details of author, title, publication, date, etc. must be incorporated within your final work and it is often a good idea to make reference to this secondary material in your text by the use of footnotes or brackets.

Remember that when you use secondary information, there is a distinction between plagiarism and the acquisition of material by research. The distinction lies in the use made by you, the candidate, of the information that you have obtained and the acknowledgement of sources used. If you make a direct quotation, then you must use quotation marks, and you should also ensure that it is properly referenced. This is why the examination boards insist that you sign a declaration of authenticity.

Before you start an investigation, check that the sources you are intending to access have the information that you need and that it is in a form you want. It will make life very difficult if, at an advanced stage of your investigation, you find that your secondary sources do not match up with your primary research. A good example occurs in crime surveys. You may have carried out the primary research on an individual street basis, but when you come to access the secondary data, in this case urban

crime figures, you will find that they are only available for each *district* within the urban area. It is therefore impossible to match them up with your more detailed figures.

Sources of secondary material

▶ **National government material** covers a wide range of data such as those concerned with the economy, employment, population and crime. Material is published by the Office of National Statistics and is available through Her Majesty's Stationery Office (HMSO), although this material can be expensive to buy and is best accessed through your local library or on the internet. The most helpful publications will probably be:
  - *Annual Abstract of Statistics*, which covers a wide range of data and includes material for previous years
  - *Population Trends*
  - *Social Trends*
  - *Economic Trends*
  - *Monthly Digest of Statistics*, the best source of up-to-date information

You could also contact government agencies for information.

▶ **Other national sources** include:
  - the media (newspapers, magazines and television/radio programmes)
  - charities
  - national organisations and action groups such as Shelter, English Heritage, The National Trust, The Countryside Commission
  - environmental pressure groups such as Greenpeace and Friends of the Earth
  - national company publications
  - the Meteorological Office

▶ **Local data** can be obtained from sources such as:
  - the local authority
  - the electoral register
  - your local library, which will have population statistics for areas as small as electoral wards (*Small Area Statistics*), census material going back to the nineteenth century and back copies of local newspapers, as well as possible photographic archives and photocopying facilities
  - the local chamber of commerce
  - estate agents
  - local newspapers
  - *Yellow Pages/Thomson Local Directory*
  - the local health authority (information on births, deaths, mortality rates and material relating to living conditions such as persons per room in households)
  - local action groups

▶ **Geographical material** comes from traditional sources including:
  - geographical magazines and journals such as *Geography/Teaching Geography, Geographical Magazine, Geofile* and *Geography Review*
  - maps and charts. Map sources include Ordnance Survey, Geological Survey, local authorities and Charles Goad, whose maps show the ownership of CBD property

▶ **The internet** offers an increasing number of sites that could be helpful in the course of a geographical investigation. It is important that you do not waste your time searching in the hope of finding something useful. Have a definite purpose in mind before accessing this source. There is a great deal of information on government sites, particularly that of the Office of National Statistics (ONS).

Presenting your results

Selecting the right technique

When you come to present your results, it is important that you use appropriate techniques. There is a wide range of techniques available to you — graphical, cartographic and tabular — but what you eventually select must be appropriate for the purpose of the investigation.

Such methods are rarely used as an end in themselves, but they are a significant element in analysis. They should be selected and applied to the data to enable you to describe any changes that are present, establish any differences and identify relationships. You should never be tempted to use as many different techniques as possible, as this can lead to similar data being presented in several different ways for no other reason than to show that you know how to construct various forms of maps and diagrams.

At this level, mark schemes award credit to those candidates who use a suitable range of techniques that can provide the potential for analysis. Most investigations will include a choice of techniques along the following lines:
▶ identification or description of differences
▶ description of spatial patterns
▶ identification of relationships
▶ classification of data according to characteristics

Whatever technique you eventually choose as being the most appropriate for the circumstances, make sure that it is clear and easy to understand, that it is as simple as possible, and that it helps you convey the message to the person reading your report.

Main methods of presentation

Use	Graphical	Cartographic
Identifying differences	Line graphs (arithmetic and logarithmic) Cumulative frequency curves (including Lorenz curves) Pie/bar graphs Proportional symbols Histograms Long/cross-sections	Pie graphs, bar graphs and proportional symbols can be placed on a base map to show spatial variations
Describing spatial patterns		Isopleths Choropleths Flow diagrams and desire lines
Identification of relationships	Scattergraphs	
Classification of data	Triangular graphs	

Arithmetic graphs

Arithmetic graphs are appropriate when you want to show *absolute changes* in the data. You will already be familiar with the use of such graphs, but there are a number of points of which you should be aware if it is your intention to use this method at this particular level:

▶ It is usual to plot the independent variable on the horizontal axis and the dependent variable on the vertical axis. With temporal graphs, time should always be considered the independent variable and plotted horizontally.

▶ Try to avoid awkward scales and remember that the scale you choose should enable you to plot the full range of data for each variable.

▶ Axes should always be labelled clearly.

▶ If you are plotting more than one line, it is a good idea to use different symbols for the plots.

▶ You can put two sets of data on the same graph, using the two vertical axes to show different scales.

Logarithmic graphs

A logarithmic graph (see Figure 1 for an example) is drawn as an *arithmetic* line graph except that the scales are divided into a number of cycles, each representing a ten-fold increase in the range of values. If the first cycle ranges from 1 to 10, the second will extend from 10 to 100, the third from

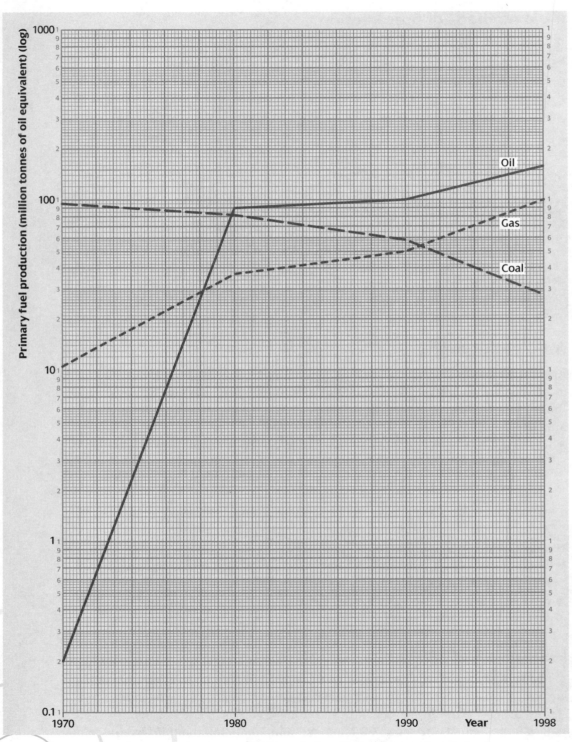

Figure 1 UK production of primary fuels (million tonnes of oil equivalent)

100 to 1000 and so on. You can start the scale at any exponent of 10, from as low as 0.0001 up to 1 million; the starting point clearly depends on the range of data to be plotted.

Graph paper can be either fully logarithmic or semi-logarithmic (where one axis consists of one or more log cycles and the other is linear or arithmetic). This is useful for plotting rates of change through time. For instance, if the rate of change is increasing at a constant, proportional rate (e.g. doubling each time period), this will appear as a straight line. In this case, you must plot time on the linear scale.

Therefore, logarithmic graphs are good for showing rates of change — the steeper the line, the faster the rate. They also allow a wider range of data to be displayed.

Remember that you cannot plot positive and negative values on the same graph and that the base line of the graph is never zero, as this is impossible to plot on such a scale.

Logarithmic and semi-logarithmic graph paper for you to photocopy is provided on pages 44 and 45.

Lorenz curves

Lorenz curves are a form of *cumulative frequency* curve. These can be drawn on both arithmetic and logarithmic axes. Data are converted into percentages and added cumulatively until 100% is reached. On the graph, each cumulative number, starting with the largest, is plotted. For example, if the highest number is 50%, it is plotted at that point. If the second highest is 20%, then that is plotted at 70% (50 + 20) and so on, until 100% is reached (see Figure 2, page 22, for an example).

A Lorenz curve can be used to measure or illustrate the extent to which a geographical distribution is even or concentrated. In this case, the largest category is taken first in the rank order, then the second and so on. The percentage of the highest category is plotted first. The second plot consists of the adding together of the highest and second highest categories. This goes on until all the categories have been plotted and 100% has been reached. This could be used, for example, if you wanted to plot the distribution of employment in an industry in relation to the workforce nationally (e.g. the distribution of all medical employment in a country against the distribution of all service employment in that country).

The location quotient, which is dealt with in the later chapter on analysis, is a method that represents such concentrations as a numerical value.

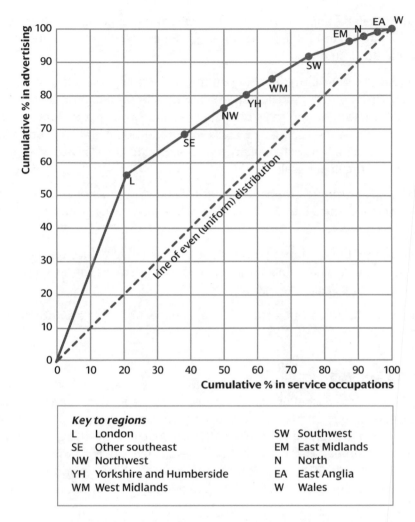

Key to regions

L	London	SW	Southwest
SE	Other southeast	EM	East Midlands
NW	Northwest	N	North
YH	Yorkshire and Humberside	EA	East Anglia
WM	West Midlands	W	Wales

Figure 2 A Lorenz curve to show the cumulative percentage of service workers in advertising in the standard regions of the UK in relation to the cumulative percentage of employment in all service occupations

If you do not need to compare a distribution with a national or regional one, then the cumulative percentage for one set can simply be plotted in rank order. The vertical axis is labelled **cumulative percentage** (scale 1–100), and the horizontal one **rank order** (scaled to cover the total number of items in the set, i.e. eight categories will produce eight rank orders (see, for example, Figure 3). This can allow comparisons with other distributions and a line of perfect regularity can be drawn where there is the same percentage in each category. This method is often used for showing the concentration of employment into particular occupations or for showing the dependence of a country on certain types of energy.

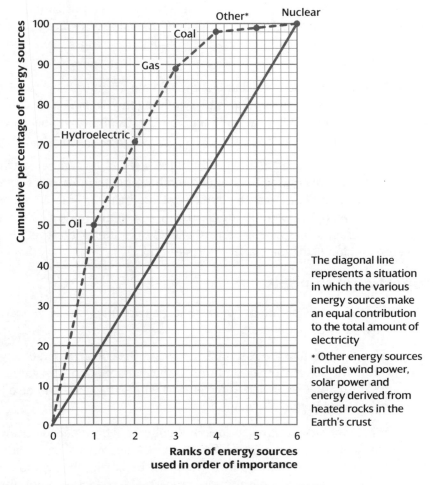

The diagonal line represents a situation in which the various energy sources make an equal contribution to the total amount of electricity

* Other energy sources include wind power, solar power and energy derived from heated rocks in the Earth's crust

Figure 3 Energy sources used to generate electricity in Italy (1995)

Pie graphs (or pie charts/divided circles) and bar graphs

In both of these methods, an area is drawn so that its size is *proportional to the value that it represents*. Circles or bars can then be sub-divided so that the components that make up each can be seen.

The pie graph is divided into segments according to the share of the total value represented by that segment. This is a useful method and visually effective as the reader is able to see the relative contribution of each segment to the whole. On the other hand, it is difficult to assess percentages if there are a lot of segments and many of them are small. Comparisons are often difficult between individual pie charts, particularly where many small segments are involved.

The bar graph (or chart) has vertical columns rising from a horizontal base, the height of the column being proportional to the value it represents. Such a vertical scale can be used to represent absolute data or those expressed in percentages.

Bar graphs are easily understood and show relative magnitudes very effectively. It is also possible to show positive and negative values if a scale is drawn through zero. In this way, profit and loss, for example, can be shown on the same graph.

Bar graphs are easy to read as the height of the bar can be compared with the vertical scale. This makes them much more useful, in most cases, than pie graphs. Many students, in their coursework, produce bar graphs that are far too complicated (for example, constructing too many multiple bars or using two different scales on the vertical axis), losing the greatest assets of this method — simplicity and clarity of presentation.

Proportional symbols

With this method, *symbols* are drawn so that they are *proportional in area or volume to the value they represent.*

We have already seen that the essential element in the construction of a bar graph is that the length of the column is proportional to the value it represents. This can be extended to pie graphs where the circle can also be drawn proportional to the total value.

Other symbols include squares, cubes and spheres. These can be drawn independently or placed on a map to show spatial differences. In the latter case, it is very important that you take great care in placing the symbols on the map. It is essential to avoid too much overlap, but each symbol must 'have a sense of place' in that it must be clear which area the symbol is representing. In AS/A2 coursework, the most commonly used methods are the bar and the circle, as these are the easiest to construct.

Histograms

This method is used to show the frequency distribution of data. It uses bars to indicate the frequency of each class of the data but it must not be confused with the bar graph/chart. Histograms are used when you want to simplify and clarify data that are easier to analyse when placed into groups, or classes, rather than being presented as individual data. By this method, you will reduce large amounts of data to more manageable proportions which will allow you to see some of the trends present within them.

Before you can draw the histogram, you will need to group the data and this can be difficult. The important principle is that you need to illustrate

differences between classes while keeping the variation within each class to an absolute minimum. You will need to establish:

▶ the number of classes to be used
▶ the range of values in each class, i.e. the class interval

The number of classes that you use must depend on the amount of data you have collected. Choose too many and you will have insufficient variation between them and may finish up with too many 'empty' classes; choose too few and you will often have difficulty in recognising trends within the data. One method is to use the formula:

number of classes = 5 × log of the total number of items in the set

If, for example, you had collected data about the size of 120 pebbles on a beach, the maximum number of classes would be:

$$5 \times \log 120 = 5 \times 2.08 = 10.4$$

You would therefore select 10 classes.

The range of values is clearly influenced by the number of classes that you have decided to use. This is shown by the formula:

$$\text{class interval} = \frac{\text{range of values (highest to lowest)}}{\text{number of classes}}$$

If, for example, you had data ranging from 96 to 5 and you required 4 classes, the class interval would be:

$$\frac{96 - 5}{4} = 22.75$$

It is important that class boundaries are well defined so that all individual pieces of data can be assigned without problem. Class intervals of 0–25, 25–50, 50–75, 75–100 should therefore be replaced with 0–24.9, 25–49.9, 50–74.9, 75–100, which are now continuous classes with no overlap.

The decision regarding the number of classes and the interval should be influenced by the type of data with which you are dealing and the purpose to which they are being put. You will need to decide exactly what it is that you are trying to illustrate or analyse. The distributions that you finish up with can be described in one of three ways:

▶ If your distribution has a modal class in the middle with progressively smaller bars to each side, then it is similar to the normal distribution (Figure 4(a)).
▶ If the modal class lies in the lower classes, then the distribution is said to show positive skew (Figure 4(b)).
▶ If it lies to the upper end, the distribution is said to show negative skew (Figure 4(c)).

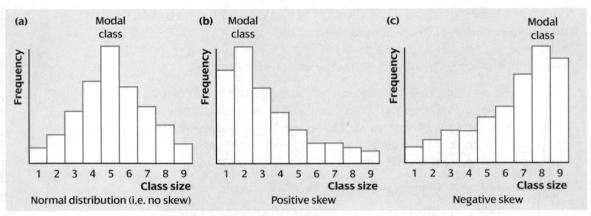

Figure 4 *Positive and negative skew*

Triangular graphs

Triangular graphs are plotted on special paper which is in the form of an equilateral triangle. Although this looks, on the surface, to be a method that has widespread application, it is only possible to use it for a whole figure that can be broken down into three components expressed as percentages. The triangular graph therefore cannot be used for absolute data or for any figures that cannot be broken down into those three components. (See Figure 5 for an example.)

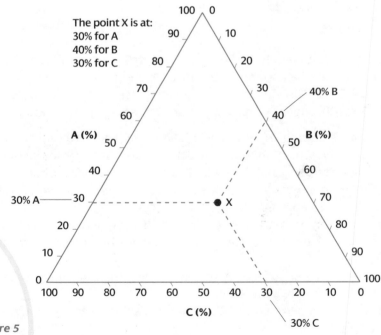

Figure 5

The advantage of using this type of graph is that the varying proportions can be seen, indicating the relative importance of each. It is also possible to see the dominant variable of the three. After plotting, clusters will sometimes emerge, enabling a classification of the items involved. They could be, for example, MEDCs/LEDCs or types of soil (see Figure 6). Triangular graph paper for you to photocopy is provided on page 46.

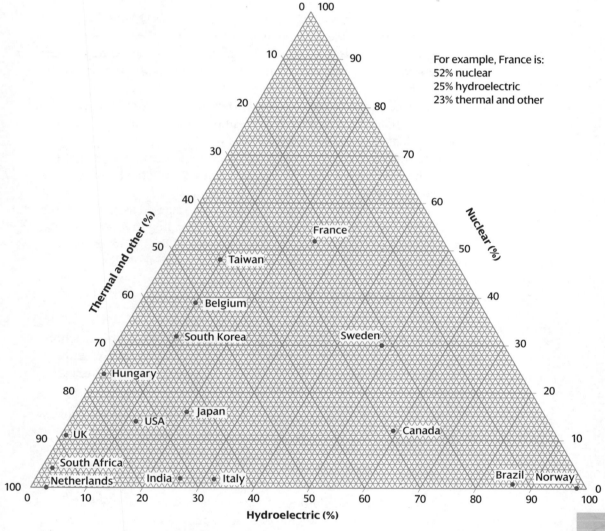

For example, France is:
52% nuclear
25% hydroelectric
23% thermal and other

Figure 6 Percentage of total electricity production by generating source for selected countries (1988)

Scattergraphs

Scattergraphs are used to investigate the *relationship between two sets of data.* In this book they are included as a form of presentation, but they are equally useful in identifying patterns and trends which might lead to further enquiry. If you use this method, there are several points to bear in mind:

▶ Scattergraphs can be plotted on arithmetic, logarithmic or semi-logarithmic graph paper.

▶ Only plot when you feel that there is a relationship to be investigated. It is possible for a correlation to emerge even when a relationship might just be coincidental.

▶ One variable usually has an effect on the other and this should enable you to identify the independent and dependent variables.

▶ The two variables are placed on the graph so that the **independent variable** goes on the *horizontal* axis and the **dependent variable** on the *vertical* axis.

▶ A trend line can then be inserted (best fit). If this runs from bottom-left to top-right, it indicates a **positive relationship**. If it runs top-left to bottom-right, it indicates a **negative relationship**.

▶ The closer the points are to the trend line the greater the degree of the relationship, but this should be assessed with a statistical test (see the Spearman rank correlation coefficient, page 33).

▶ Points lying some distance from the trend line are classed as residuals (the anomalies). These can be referred to as either negative or positive. Identification of residuals can enable you to make further investigations into the other factors that could influence your selected two variables.

Long/cross sections

These methods are useful for describing and comparing the shape of the land. Long sections are mostly used with river studies, while cross sections can be drawn for a number of landscape features. This method essentially consists of constructing a graph that shows height against a horizontal scale of distance.

Horizontal scale is usually taken as that on the map with which you are working, but it is very unusual for you to use the same vertical scale. If working with a standard OS map, this would reduce your section, in most cases, to a line showing little variation. It is therefore necessary to adopt a larger scale, but care must be taken to ensure that it is not massively exaggerated and that the gentlest of valley slopes is not changed into the north wall of the Eiger! The degree of exaggeration can be calculated and presented with your finished work. The same principles apply to both long and cross section construction.

Isopleth/isoline maps

When you collect data that can be represented as points on a map, it is possible to draw a map where all points of the same value are joined by a line. Such a map allows a pattern to be seen in a distribution. The best known example is on OS maps where the lines (contours) join places of the same height. This technique can be applied to a number of other physical factors, such as rainfall (isohyets), temperature (isotherms) and pressure (isobars), as well as a number of human factors, such as pedestrian densities in a CBD or travel times (isochrones) for commuters and shoppers.

Choropleth maps

A choropleth is a map on which the density of shading within each area shown is proportional to the value worked out for that area. Differences in shading therefore indicate the changes between areas. The data are usually in a form that can be expressed in terms of the area, such as population density per square kilometre.

To produce such a map, there are certain stages that have to be followed:
▶ The material has to be grouped into classes. Therefore you need to decide the number of classes required to display the full range of your data and suitable class boundaries which give you the range for each class.
▶ You need to devise a range of shadings to cover the range of your data. Darkest shades should represent the highest figures, and vice-versa. It is good practice not to use the two extremes of black and white; black suggests a maximum level, while white indicates that there is nothing within the area. A suitable method of shading is shown in Figure 7.

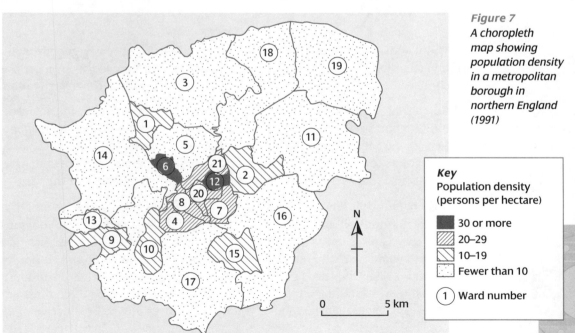

Figure 7
A choropleth map showing population density in a metropolitan borough in northern England (1991)

Key
Population density (persons per hectare)

■ 30 or more
▨ 20–29
▧ 10–19
⬚ Fewer than 10

① Ward number

Such maps are fairly easy to construct and are visually very effective as they give the reader a chance to see general patterns in an areal distribution. There are, however, a few limitations to the method:

▶ It assumes that the whole area under one form of shading has the same density, with no variations. For example, on maps of the UK the whole of Scotland may be covered by one category, when it is obvious that there could be large variations between the central populated areas and the Scottish Highlands.

▶ The method indicates abrupt changes at the drawn boundaries which will not be present in reality.

Flow lines/desire lines

Flow diagrams are drawn to show the quantity of movement that follows an actual route. In this case, each route is drawn to a width that represents the amount of traffic etc. moving along it. The width is therefore proportional to the quantity of movement. Routes can sometimes be drawn in a straight line linking an origin and a destination — shopping movements, for example. These are also known as desire lines.

The analysis and interpretation of data

The use of statistical analysis is a common feature of many geographical investigations. Such objective analysis of data can be used to support the conclusions suggested by a subjective view of the results of the investigation.

Statistical analysis should not be used just for show. It should form an integral part of the coursework write-up. There should be careful consideration of the most effective form of statistical analysis, together with some thought as to why that technique is appropriate. The best form of statistical analysis should seek to assist in evaluation of the significance of the results.

It is obvious that any statistical technique should be used correctly, and that all calculations should be performed accurately. It is perhaps less obvious that the result of any calculation should be supported by statements that explain what the mathematical result means. Many calculator and computer functions will complete the mathematical process for you, so it is paramount that you understand the relevance of the values produced by them. In short, never use a statistical technique you do not understand or in which you are not confident.

The following table summarises some of the major statistical techniques that are either required or suggested by the AS/A2 specifications of the major examination boards for use in coursework and/or written alternative examinations.

Reason for using statistics	Statistical technique(s)
The summarising and comparison of data	Measures of central tendency: mean, mode, median
The dispersion and variability of data	Range Inter-quartile range Standard deviation
The correlation of two sets of data	Spearman rank correlation coefficient and tests of significance
The degree of concentration of geographical phenomena	Location quotients
The measurement of patterns in a distribution	Nearest neighbour statistic
The degree to which there are differences between observed data and expected data, and the statistical significance of them	Chi-squared test

The following sections set out the means by which each of these techniques can be used, and the rationale for their use.

Measures of central tendency

There are three measures: the arithmetic mean, the mode and the median.

Arithmetic mean ($\bar{x}$)

This is calculated by adding up all of the values in a data set and dividing the total by the number of values in the data set. So,

$$\bar{x} = \frac{\Sigma x}{n}$$

The arithmetic mean is of little value on its own, and should be supported by reference to the standard deviation of the data set.

Mode

This is the most frequently occurring value in a data set, and can only be identified if all the individual values are known.

Median

This is the middle value in a data set when arranged in order of highest to lowest (that is, in rank order). There should be an equal number of values both above and below the median value. If the number of values in a data set is odd, then the median will be the $\frac{(n+1)}{2}$ th item in the data set. So, for example, if the total number of items in a data set is 27, then the median will be the 14th value in the rank order of the data.

If the number of values in the data set is even, then the median value is the mean of the middle two values. Any calculation of the median is best supported by a statement of the interquartile range of the data.

It is possible that each of these measures could give the same result, but they are more likely to give different results. The characteristics of the distribution of a data set would have to be perfectly normal for the same result to be provided, and this is extremely unlikely when using real data. It is more likely that the distribution of the data set is skewed. The more it is skewed, then the greater the variation in the three measures of central tendency.

None of these measures gives a reliable picture of the distribution of the data set. It is possible for two different sets of data to give the same values for mean, mode and median. Consequently, measures of the dispersion or variability of the data should also be provided.

Measures of dispersion or variability

There are three measures of dispersion or variability: range, inter-quartile range and standard deviation.

Range
This is the difference between the highest value and the lowest value in a data set. It gives a simple indication of the spread of the data.

Inter-quartile range
This is the difference between the upper quartile and the lower quartile of a data set.
▶ The upper quartile (UQ) is the $\frac{(n+1)}{4}$ th item in the data set when arranged in rank order (from highest to lowest).
▶ The lower quartile (LQ) is $3\frac{(n+1)}{4}$ th item in the data set when arranged in rank order.

So the inter-quartile range (IQR) = UQ − LQ.

The IQR indicates the spread of the middle 50% of the data set about the median value, and thus gives a better indication of the degree to which the data are spread, or dispersed, on either side of the middle value.

Standard deviation
This measures the degree of dispersion about the mean value of a data set. To calculate it, the following steps should be taken:
▶ The difference between each value in the data set and the mean value should be calculated.

▶ Each difference should then be squared, to eliminate negative values.
▶ These squared differences should be totalled.
▶ This total should be divided by the number of values in the data set, to provide the variance of the data.
▶ The square root of the variance should be calculated.

$$\text{standard deviation} = \sqrt{\frac{\sum (\bar{x} - x)^2}{n}}$$

The standard deviation is statistically important as it links the data set to the normal distribution. In a normal distribution:

▶ 68% of the values in a data set lie within ±1 standard deviation of the mean
▶ 95% of the values in a data set lie within ±2 standard deviations of the mean
▶ 99% of the values in a data set lie within ±3 standard deviations of the mean

A low standard deviation indicates that the data are clustered around the mean value and that dispersion is narrow. A high standard deviation indicates that the data are more widely spread and that dispersion is large. The standard deviation also allows comparison of the distribution of the values in a data set with a theoretical norm and is therefore of greater use than just the measures of central tendency.

Measuring correlation: the Spearman rank correlation coefficient

Comparisons are made between two sets of data in order to infer a relationship between them. However, it must be noted that, even if there is a relationship, it does not prove a causal link. In other words, the relationship does not prove that a change in one variable has been responsible for a change in the other. For example, there may be a direct relationship between altitude and the amount of precipitation in a country such as the UK. These two variables (altitude and precipitation) are clearly linked, but a decrease in one does not automatically cause a decrease in the other — they are simply related to each other. There are two main ways in which comparisons can be shown: one is by scattergraphs (see page 27); the other is the measurement of correlation by the Spearman rank correlation coefficient.

The latter is used to measure the degree to which there is correlation between two sets of data (or variables). It provides a numerical value in order to summarise the degree of correlation, and hence it is an example of an objective indicator. Once calculated, the numerical value has to be tested statistically to see how significant the result is.

The test can be used with any sets of data consisting of raw figures, percentages or indices that can be ranked. The formula for the calculation of the correlation coefficient is:

$$R_s = 1 - \frac{6\Sigma d^2}{n^3 - n}$$

where d is the difference in ranking of the two sets of paired data and n is the number of sets of paired data.

The method of calculation is as follows:
▶ Rank one set of data from highest to lowest (highest value ranked 1, second highest 2 and so on).
▶ Rank the other set of data in the same way.
▶ Beware of tied ranks. In order to allocate a rank order to such values, calculate the average rank that they occupy. For example, if there are 3 values all of which should be placed at rank 5, then add together the ranks 5, 6 and 7 and divide by 3, giving an average rank of 6 for each one. The next value in the sequence would then be allocated rank 8.
▶ Calculate the difference in rank (d) for each pair of data.
▶ Square each difference.
▶ Add the squared differences together and multiply by 6 (step A).
▶ Calculate the value of $n^3 - n$ (step B).
▶ Divide step A by step B, and take the result away from 1.

The answer should give you a value between +1.0 (perfect positive correlation) and −1.0 (perfect negative correlation).

Some words of warning
▶ You should have at least 10 sets of paired data, as the test is unreliable if n is less than 10.
▶ You should not use too many sets of paired data (maximum 30?), as the calculations become very complex and prone to error.
▶ Too many tied ranks can interfere with the statistical validity of the exercise, although it is appreciated that there is little you can do about the real data collected.
▶ Be careful about choosing the variables to compare − do not choose dubious or spurious sets of data.

How to interpret the result of a Spearman rank correlation
In trying to interpret the result of the Spearman rank correlation test the following should be considered:

What is the direction of the relationship?
If the calculation produces a positive value, then the relationship is positive, or direct. In other words, as one variable increases, then so does the other. If

the calculation produces a negative value, then the relationship is negative, or inverse.

How statistically significant is the result?

When comparing two sets of data, there is always a possibility that the relationship shown between them has occurred by chance. The figures in the data sets may just happen to have been the right ones to bring about a correlation. It is necessary, therefore, to have a method by which the statistical significance of the result can be assessed. In the case of the Spearman rank correlation coefficient test, the critical values for R_S must be consulted. These can be obtained from statistical tables, but the table below shows some examples of them.

n	0.05 (5%) significance level	0.01 (1%) significance level
10	± 0.564	± 0.746
12	0.506	0.712
14	0.456	0.645
16	0.425	0.601
18	0.399	0.564
20	0.377	0.534
22	0.359	0.508
24	0.343	0.485
26	0.329	0.465
28	0.317	0.448
30	0.306	0.432

According to statisticians, if there is a > 5% possibility of the relationship occurring by chance, then the relationship is not significant. This is called the rejection level. The relationship could have occurred by chance more than five times in a hundred, and this is an unacceptable level of chance. If there is a < 5% possibility, then the relationship is significant, and therefore meaningful. If there is a < 1% possibility of the relationship occurring by chance then the relationship is very significant. In this case, the result could only have occurred by chance 1 in 100 times, which is very unlikely.

So how does this work? Having calculated a correlation coefficient, examine the critical values given above (ignore the positive or negative sign). If your coefficient is greater than these values, then the coefficient is significant at that level. If your coefficient is smaller, then the relationship is not significant at that level.

As an illustration, suppose you had calculated an R_S value of 0.50 from 18 sets of paired data. 0.50 is greater than the critical value at the 0.05 (5%) level, but not that of the 0.01 (1%) level. In this case, therefore, the relationship is significant at the 0.05 (5%) level, but not at the 0.01 (1%) level.

Location quotients

A location quotient is a measure of the degree to which a geographical activity is concentrated in an area. As in the case of the Spearman rank correlation coefficient, the end product is a number, which again gives an objective value with which to compare. The numerical value compares the concentration of an activity in a sub-region with the concentration in the whole region.

For example, when studying the concentration of employment in an industry in a particular region of a country, the calculation of the location quotient is

$$LQ = \frac{X'/Y'}{X/Y}$$

where:
- X' is the number employed in the given industry in the region
- Y' is the number employed in all industries in that region
- X is the number employed in the given industry in the country as a whole
- Y is the number employed in all industries in that country

In this case, the LQ compares the proportion of employed people in a particular region in a given industry with the proportion in that industry nationally.

Sometimes data are given in percentage form, in which case the calculation is more straightforward:

LQ = (% of workers in the given activity in the region) divided by (% of workers in the activity nationally)

The key indicator in the use of location quotients is LQ = 1.0. A result of this nature would indicate that the region in question has a fair share of that geographical activity compared with the rest of the country. The relative proportions, region vs country, are equal.

If LQ > 1.0, then this would indicate a greater share of that activity in that region compared with the rest of the country. In other words, there is a concentration of that activity in that region.

Similarly, if LQ < 1.0, then the region has less that its proportionate share — it is under-represented.

The nearest neighbour statistic (R_n)

This statistic analyses the distribution of individual points in a pattern. It can be applied to the distribution of any items that can be plotted as point locations. Consequently, it is often used to analyse the distribution of certain shop types in a town centre, the distribution of various sizes of settlements in an area, and the distribution of some public services, for example doctors' surgeries, in an urban area.

The basis of the statistic is the measurement of the distance between each point in a pattern and its nearest neighbour. This must be done for each point identified within the area to be studied. Once all measurements have been completed, the mean distance ($\bar{d}$) between each pair of points should be calculated.

The nearest neighbour statistic can then be calculated using

$$R_n = 2\bar{d}\sqrt{\frac{N}{A}}$$

where:
▶ N is the number of point locations in the area
▶ A is the area of study

The statistic can be any value between 0 and 2.15.

0 represents a pattern that is perfectly clustered, that is, there is no distance between nearest neighbours — all the points are at the same location. If this were the case, there would be no pattern to analyse.

2.15 represents a pattern displaying perfect regularity — all points lie at the vertices of equilateral triangles, forming a perfect pattern. All distances between nearest neighbours are identical. Again, this is highly unlikely in the real world.

1.0 is said to represent a random pattern, although this is difficult to prove.

In practice, the outcome of such a calculation will be on the continuum between 0 and 2.15. Proximity to one of these will indicate the degree of either clustering or regularity.

When carrying out a nearest neighbour statistic calculation, it is advisable to map the activity(ies) first onto a transparent overlay. This removes any potential distraction in the measurement and calculation process. Another necessary requirement is that the units of measurement of distance and of area are the same, in most cases either metres and square metres, or kilometres and square kilometres.

A further complication is the delineation of the area to be studied. In the case of settlement-based work, you have to decide on the boundary of the area you are studying. You can then establish a buffer zone around this study area. When measuring the distance between each point and its nearest neighbour, if the nearest neighbour is within this buffer zone then you should measure *to* this point. However, this point in the buffer zone should not be counted in the overall study as a point from which to measure to its nearest neighbour. In short, you should only measure *to* points in the buffer zone, not *from* them.

The chi-squared test

This technique is used to assess the degree to which there are differences between a set of collected (or observed data) and a theoretical set of data (or expected data), together with the statistical significance of them.

The observed data are those that have been collected either in the field or from secondary sources. The expected data are those that would be expected according to the theoretical hypothesis being considered or being tested.

Some words of warning

This test is a difficult concept to grasp fully and should not be attempted unless you are confident in your ability both to apply it and to understand the outcome. If neither of these is the case, then you should move on to the next section now.

Normally, before the test is applied, it is necessary to formulate a null hypothesis. In this case, the null hypothesis would be that there is no significant difference between the observed and expected data distribution. The alternative to this would be that there is a difference between the observed and expected data, and that there is some factor responsible for it occurring.

The method of calculating chi-squared is shown below. The letters A to D in the table refer to map areas A to D drawn above the table (Figure 8). In the column headed O are listed the numbers of points in each of the areas A to D on the map (the observed frequencies). The total number of points in this case is 40. Column E contains the list of expected frequencies in each of the areas A to D, assuming that the points are evenly spaced. In the column $O - E$, each of the expected frequencies is subtracted from the observed frequencies, while in the last column the result is squared. The relevant values are then inserted into the expression for chi-squared, and the resultant value is 3.2.

Figure 8

Map	Observed (O)	Expected (E)	O - E	(O - E)²
A	8	10	-2	4
B	14	10	4	16
C	6	10	-4	16
D	12	10	2	4
Sum	**40**	**40**	**0**	**40**

$$\chi^2 = \sum \frac{(O-E)^2}{E}$$

$$= \frac{40}{10}$$

$$= 4.0$$

The aim of a chi-squared test, therefore, is to find out whether the observed pattern agrees with or differs from the theoretical (expected) pattern. This can be measured by comparing the calculated result of the test with its level of significance.

To do this, the number of degrees of freedom must first be determined. This is done using the formula ($n - 1$), where n is the number of observations, in this case the number of cells that contain observed data (4). So, $4 - 1 = 3$. Statistical tables give the distribution of chi-squared values for these degrees of freedom.

Then there are the levels of significance. There are two levels of significance: 95% and 99%. At 95%, there is a 1 in 20 probability that the pattern being considered occurred by chance, and at 99% there is only a 1 in 100 probability that the pattern is a chance one. The levels of significance can be found in a book of statistical tables. These levels of significance are also called confidence levels.

If the calculated value is the same as or greater than the values given in the table, then the null hypothesis can be rejected and the alternative hypothesis accepted.

In the case of our example, however, the value of chi-squared is very low (3.2), showing that there is little difference between the observed and the expected pattern. The null hypothesis cannot therefore be rejected.

Some further points on this technique
▶ The numbers used in both the observed and expected values column must be large enough to ensure that the test is valid. Most experts state that there should be a minimum of 5.
▶ The number produced by the calculation is itself meaningless. It is only of value when used in consultation with statistical tables.

▶ Only significance (or confidence) levels of 95% and 99% should be considered (as in the case of Spearman rank) when rejecting the null hypothesis. Any levels of confidence greater than these simply allow the null hypothesis to be rejected with even greater confidence.

▶ It is strongly recommended that you do not apply the test with more than one set of observed data. In this case, the mathematics become too complex.

And finally...

It is important to note that all of these techniques are only to be used to support your own ideas on the geographical significance of your study. All results, and the statistical analysis of them, should be related to the original hypothesis and/or the established theory in that aspect of the subject.

Your results may support established theory or your hypothesis, or they may not. If the latter, there may be some reason or factor that is responsible which could then lead to further studies.

Above all, your project or coursework should make geographical sense. This is far more important than demonstrating your ability to use mathematics or statistics.

Writing up the report/coursework

The final write up of your enquiry should be well structured, logically organised, and clearly and concisely written. There are three aspects of this process that you should consider: structure, language and presentation.

Structure

The structure should assist the reader in understanding the report, and should also assist you in organising it logically. The following check list provides a generalised structure to your report, but you should also check whether your specification requires particular sections in addition to these:

▶ report approval sheet/cover sheet
▶ title page and contents page
▶ executive summary (if required)
▶ aims and objectives
▶ scene setting (if required)
▶ research questions/hypotheses/issues being examined
▶ sources of information used
▶ methods of data collection and commentary on their limitations

▶ data presentation, analysis and interpretation
▶ evaluation and conclusion
▶ bibliography and appendices

You need not attempt these in the order given. Indeed, it may be easier if you do not. For example, the executive summary is perhaps best written at the end of the whole process, as it is only at this stage that the whole picture can be described. What follows is a suggested order of completion.

(1) Data presentation, analysis and interpretation

This is the section in which you present and analyse your findings. At this stage you will have collected the data, and have sorted them and selected the most useful pieces. You will know what you have found out, and what it all means. Your results will be complete, and they will be most fresh in your mind at this time. You should be able to interpret each separate section of your results and formulate conclusions for each one. The whole picture may begin to appear in your head.

(2) Sources of information, and methods of data collection

Now you can write about what information you collected, and how you did so. Don't forget to discuss any limitations of the methods of collection that you used, or of the data sources themselves.

(3) Conclusion and evaluation

This should include a summary of all the major findings of your enquiry and an evaluation of them. Do not present anything new to the reader at this stage. Towards the end of this section, try to draw together each of the sub-conclusions from each section of the data analysis into one overall conclusion — the whole picture.

(4) Aims, objectives, research questions and scene setting (the introduction)

Having written up the bulk of the enquiry, you can now write the introduction, making sure it ties in with what follows it. The purpose of this section is to acquaint the reader with the purpose of the enquiry, and the background to it.

(5) Appendices and bibliography

The appendices comprise additional pieces of evidence that may be of interest to the reader, but are not essential to the main findings. The bibliography provides detail of the secondary sources that have been used in your research, either as guides or as sources of information. You should give the author, the date of publication, the title of the publication and the page number. When quotes are used in the body of the report of your enquiry, you should provide the name of the author and date of publication in an appropriate place in the text, and cross-refer it back to the bibliography.

(6) Contents page

All the sections of the report should be listed in sequence, with accurate page references.

(7) Title page

This should state the title of your report, in the style required by the specification. Also add your name, candidate number, centre number and date of completion.

(8) Executive summary

An executive summary should provide a brief statement of the complete enquiry. It should be no more than 250 words in length and cover all the main aspects of the enquiry. A good executive summary introduces the subject of the full report, refers to its aims and objectives, and provides a brief synopsis of the findings. A very good executive report will tempt the reader into wanting to read more by being comprehensible, interesting and stimulating. It should also make sense and read as a separate document from the full report.

Language and length

The quality of language that you use in the writing up of your enquiry is important. You are entirely in control of this aspect of the process, and your style of writing must be appropriate for this exercise. You should avoid poor use of the English language, and try to maintain accuracy at all times. In particular:

▶ your sentences should be grammatically correct and well punctuated
▶ your writing should be well-structured, with good use made of paragraphing
▶ your spelling must be accurate (use a dictionary or spellcheck on a PC)
▶ be clear in your use of specialist terminology, and in the expression of your ideas
▶ be aware that the assessment of your work will take into account the above aspects of your writing

Proof-reading is an important part of this process. Prior to submission, make sure you read through the draft from start to finish, and mark any places where there are errors or inconsistencies. If possible, get someone to do this for you too; parents are often useful in this respect. However, you do need someone who is going to be highly critical of what you have written. A report littered with spelling mistakes or grammatical errors does not impress, and your PC spell check will not pick up all of these.

Another key checking stage is making sure that maps and diagrams are located in the correct place — it is irritating for the reader to have to flick

backwards and forwards when trying to read a report of this type. Also, make sure all the references in the text are included in the bibliography.

Finally, a word on length is necessary. All the examination boards state the required word limit for reports, and they will impose a penalty of some sort for over-long submissions. Once again, this is a reason for careful scrutiny of the requirements of your particular specification. In such cases, it is the final part of your report which is most under threat — that dealing with the conclusion and evaluation. These are important areas of your coursework and it would be foolish to make the marks available for this section inaccessible.

Word limits are not designed to make life difficult — they tend to have the opposite effect. Reports written within the stated word limit tend to be those that are better planned, structured and executed. No penalty is imposed for reports that are too short, but they are self-penalising because they contain inadequate material.

Presentation

It is a fact of life that most people are influenced by presentation, and that includes coursework examiners. Bear in mind the following:

▶ A neatly presented, handwritten or word-processed report is going to create a favourable impression before its contents are read.
▶ Adequate heading and numbering of pages, with carefully produced illustrations, will make it easier for the reader to understand what is contained within the report.
▶ Layout is also important. Do not crowd the pages with dense text, which looks unattractive. Provide adequate margins, use either double or 1.5 line spacing if using a word-processing package, and make use of clear heading levels, with short paragraphs.
▶ Check on the type of folder that is acceptable to the examination board for enclosing your work.
▶ Make sure you allow enough time to add the finishing touches in order to give your work the final polish. It goes without saying that this time will be available if you have not left the completion too near to the final deadline.

You should now be in a position to submit your finished product confident in the belief that it is the best you could have done.

Logarithmic graph paper

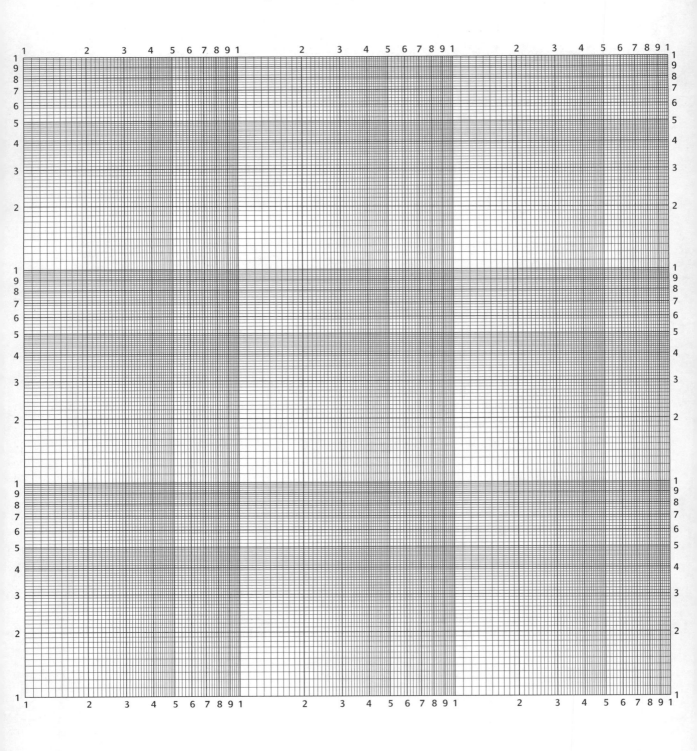

Semi-logarithmic graph paper

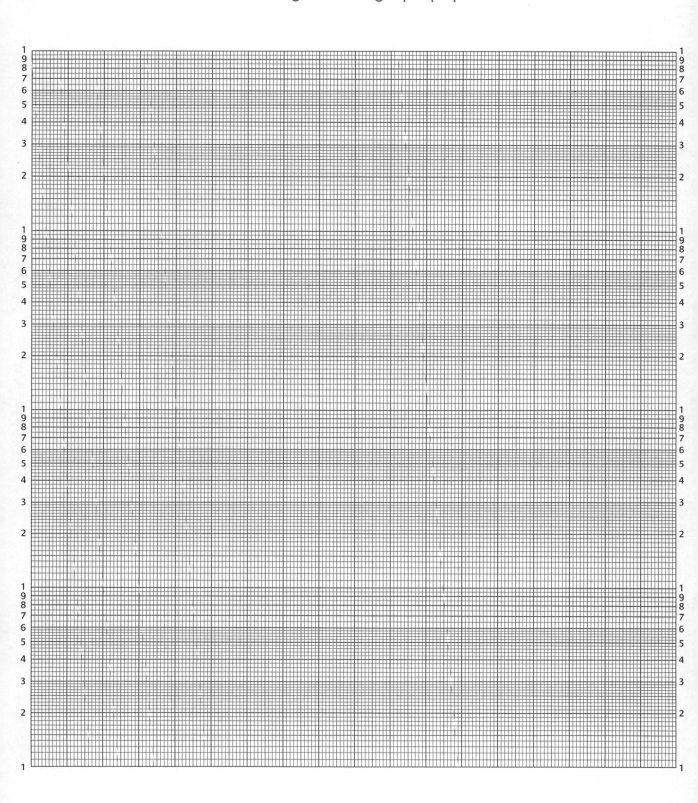

Triangular graph paper

AS/A-Level Geography

Coursework
& Practical
Techniques

Written alternative examinations

Written alternative examinations

The first section of this book examined the processes by which a geographical investigation can be completed, written up and submitted as a complete piece of coursework. However, all the GCE examination boards offer a written test as an alternative to the completion of coursework. This section examines these written alternatives by offering example questions in the style in which they will be set. Some sample answers are also provided and these are followed by examiner's comments.

Much of the material given in the first section of this book is relevant here, but this section looks at the particular needs of the written alternatives. The questions fall into two broad classes.

Firstly, some are set on a piece of fieldwork that has been undertaken by a candidate, or are based on material given to the candidate. The major difference here is that the fieldwork undertaken by the candidate (or on behalf of the candidate) has not been formally written up and submitted as a piece of assessed coursework. These questions we shall call **fieldwork-based questions**. Examples occur in:
▶ AQA Specification A Unit 7
▶ AQA Specification B Unit 6, Question 1
▶ Edexcel Specification A Unit 3, Section B
▶ OCR Specification A Units 2682 and 2686

Secondly, the written alternative papers also test fieldwork-related, or investigative, skills. Often they are based on data provided by the boards, and candidates are asked to present, analyse and evaluate those data. In other words, these questions assess the candidate's practical abilities, and we shall call them **practical questions**. Examples occur in:
▶ AQA Specification A Unit 7
▶ AQA Specification B Unit 6, Question 2
▶ Edexcel Specification A Unit 3, Section A
▶ OCR Specification A Units 2682 and 2686

What follows are examples of these two types of question, for each of the examination boards given above. As AQA Specification A Unit 7 contains both types of question, within the same context, this unit test is looked at first, but thereafter the questions are separated according to type.

Fieldwork-based/practical — AQA (A) Unit 7

AQA (A) Units 6 and 7 require you to undertake investigative work in the field to develop skills associated with planning, collection of primary and/or secondary data, presentation, interpretation and evaluation, in order to be able to produce fieldwork investigations. Unit 6 requires you to submit a 4000-word personal investigation, while for the alternative Unit 7, the assessment takes the form of a written examination. This is based on a topic published by AQA 2 years in advance of the examination and is selected from material in AS Modules 1/2 (Core Concepts in Physical/Human Geography) or A2 Modules 4/5 (Challenge and Change in the Physical/ Human Environments). These topics are always based on a small area of study.

Pre-release material

Approximately 4 weeks before the unit test, centres are sent an Information Booklet which presents material relating to the investigation in question. Within this booklet, an aim, hypothesis, question, issue or problem will be put forward, which will be explored in the written assessment unit.

The booklet contains sufficient information so that, prior to the examination, you have the chance to familiarise yourself with the purpose of the data and to assess how they can be used. Information on the method of data collection is also supplied.

You are advised to consult your own fieldwork investigative techniques that complement the hypothesis being assessed. In the written exam, you are given the opportunity to indicate your ability to apply the skills, knowledge and understanding gained in personal investigative work and to demonstrate a grasp of the issues involved. You therefore need to be familiar with the pre-release material which you can also refer to in the course of the examination.

Within the examination paper will be other material on the chosen topic, in the form of maps, diagrams, tables, photographs, etc. These will form the basis of some of the questions.

In presenting a typical Unit 7-style question in this book, it has not been possible to show to you all of the material that would have been found in both the pre-release booklet and in the examination paper. Some material, though, has been given in full, while other maps, diagrams, etc. that would have been provided have been named and their nature indicated. This clearly has a consequence for the example questions that have been set in this publication. In this case, we have produced a number of questions and answers that are representative of the types that will be set (and based upon the material that we have been able to give to you). This material, however, does not represent a whole paper; the unit test itself is much longer.

Example unit

Pre-release material

Enquiry title

> **Is it possible to define the extent of the central business district (CBD) of Fakeville?**

The basic **aim** of this enquiry is to establish if it is possible to identify the exact boundary of the CBD of Fakeville or whether it is represented as a zone of gradual change. Individual criteria used to determine the boundary will be land use, height of buildings, size of buildings, quality of buildings, pedestrian counts and rateable values.

Before beginning the enquiry, it is important to establish some **background information.** A central business district has a number of characteristics:
- it contains major retailing and financial organisations
- it has major concentrations of offices
- the tallest buildings in the city are in this area
- it has the highest land values in the city
- during the day, it has the highest pedestrian densities
- it contains the highest-quality buildings
- the area has the greatest accessibility with the greatest volume and concentration of traffic
- it is dynamic, in that it is constantly being added to in some areas but is losing parts to other functions in other areas

The booklet would also contain selected information on the population of Fakeville (size, increase, working/non-working, car ownership, manual/professional) and a location map. For example, according to the 1991 census, Fakeville had a population of 63 814; there had been an intercensal increase of 4.8%.

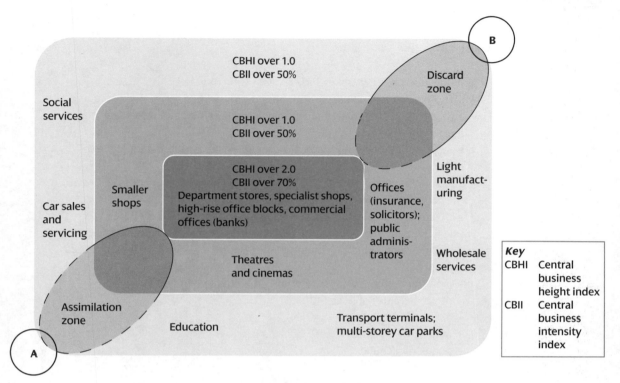

Figure 1 *Model of a typical CBD in the UK*

Data collected

The following information was collected for each building in the centre of Fakeville:

▶ ground floor function
▶ number of storeys (height)
▶ size
▶ quality/decay index (scale 1–10)

Information would also be given on how size is defined (number of front windows) and the way in which the quality/decay index score is reached. Maps would be presented showing the following:

• ground floor function of each building in the centre
• locations of pedestrian counts
• average height of buildings for each street
• average size of buildings for each street
• average quality/decay index for each street

In addition, a number of photographs would be presented that were taken in the study area and their locations shown on a map of the central area.

Question paper

The question paper would include the following material:
- a photograph taken near the CBD boundary
- a partly completed map of size and height of buildings by street, using a divided bar method
- a partly completed quality/decay index score map, using a shading technique
- a table of the pedestrian count survey, together with the rateable values at the collection points (Table 1 — supplied below)
- a table on which to complete a Spearman rank correlation coefficient for selected pedestrian counts and rateable values (Table 2 — supplied below)
- a map showing some CBD boundaries of Fakeville, based on various indicators such as
  - area covered by business rates (original area for study)
  - rateable values of over £40 000
  - function of buildings
  - quality/decay index
  - pedestrian counts

Aims

(1) With reference to your own experience of planning fieldwork enquiries and the background information in the pre-release booklet:

(a) Outline the information which may have provided the idea for this study. (5 marks)

(b) Give one reason why the CBD of Fakeville was a suitable site for this study. (1 mark)

Methods

(2) Pedestrian counts were taken at 37 predetermined positions and their locations are shown on the map. Describe the strategy which you would have adopted to select the points and comment on the relative merit of that strategy. (6 marks)

(3) The business rate data given in Table 1 are secondary data. Explain the usefulness of these data in this particular enquiry. (2 marks)

(4) Complete Table 2 in order to calculate the Spearman rank correlation coefficient for the pedestrian counts and rateable values along a transect within the CBD of Fakeville. (4 marks)

(5) Referring to Table 3 (page 54) interpret the meaning of the coefficient that you have completed in question 1. (3 marks)

Point number	Rateable value (£)	Pedestrian count	Point number	Rateable value (£)	Pedestrian count	Point number	Rateable value (£)	Pedestrian count
1	27700	16	13	20650	29	25	41500	27
2	20200	12	14	77750	15	26	66950	13
3	no data	46	15	30300	112	27	16000	2
4	5800	3	16	20850	152	28	13800	6
5	25500	2	17	63000	225	29	5150	10
6	6150	2	18	172000	306	30	12400	29
7	10850	34	19	37700	242	31	10000	37
8	25000	16	20	55000	10	32	25000	41
9	17600	22	21	997500	54	33	40800	201
10	18850	28	22	6000	55	34	57800	20
11	48700	17	23	34000	1	35	4150	7
12	90000	23	24	754000	11	36	49400	5
						37	437500	88

Table 1 Skills, techniques and interpretation

Point number	Rateable value	Rank	Pedestrian count	Rank	d	d^2
35	4150		7			
20	55000		10			
19	37700		242			
18	172000		306			
17	63000		225			
16	20850		152			
15	30300		112			
12	90000		23			
10	18850		28			
6	6150		2			

$$R_S = 1 - \left(\frac{6 \Sigma d^2}{n^3 - n} \right)$$

Σd^2

$$R_S = 1 - \left(\frac{6 \underline{\quad}}{\underline{\quad}} \right)$$

$$R_S = 1 - \left(\frac{\underline{\quad}}{\underline{\quad}} \right)$$

$$R_S = 1 - \left(\underline{\quad\quad} \right)$$

Table 2

	Significance level	
N	0.05	0.01
10	± 0.564	± 0.746

Table 3

Conclusions

(6) Write a conclusion to this enquiry with specific reference to the aims and objectives given in the pre-release booklet. Using your own experience of conducting an enquiry, you should, in addition, consider the reliability of your findings and suggest how the enquiry could be extended and improved. (10 marks)

Enquiry-related issues

(7) Within Fakeville, this enquiry has focused on the delimitation of the CBD. It would be possible to broaden the enquiry to take in other related issues:
A the impact of tourism within the CBD
B the changing sphere of influence of the CBD

Choose **A** or **B** and then:
(a) Suggest a hypothesis, question or problem that could be investigated. (2 marks)

(b) Briefly describe the study area (location and nature) where you would carry out your fieldwork, and justify your choice. (4 marks)
(c) Identify two items of data needed for your chosen enquiry. For each item, describe the way in which you would collect the data in your study area and justify the method with reference to your hypothesis. (6 marks)

Sample answers

(1) (a) CBDs have many characteristics and it is therefore possible to break them up into several zones in which these characteristics are dominant. Models of CBDs show such zones and the changes that occur as one moves across a city. It may well be that, as the model shows, such changes are very gradual, but from my experience of fieldwork, it must be possible to define where one zone begins and another ends, as some characteristics change very quickly.

There is also information on the model that could be conflicting. For instance, one could ask whether the frame really belongs to the CBD. The intensity index of 50% has been taken as the theoretical boundary, but is this really sustainable on the ground? The model, however, provides a clear basis for this enquiry, as the criteria change from the centre of the CBD outwards. With the zones of assimilation and discard, the boundaries can be seen to change over time.

e This is a very good response, for full marks. The answer is targeted to the question in that the candidate is aware of the gradual change implied by some of the data on the model, but has also noted the uncertainty that some of the information implies. The change of characteristics through both time and space has been introduced, along with some reference to the candidate's own fieldwork. It is hard to see how the candidate could have improved upon this answer.

(b) Fakeville has a population of around 60 000, resulting in the development of a CBD of a reasonable size, with recognised features.

e There is only 1 mark available here and the candidate has secured it with this answer. Another point that could have been offered is the manageable size of the CBD, within which there would be no need to use transects.

(2) I would have used a systematic method to choose my sampling points within Fakeville, resulting in equidistant counting points. This would have allowed me to cover the entire area equally and avoided any clustering which may have occurred if a random method of sampling had been used. Clustering would result in other parts of the CBD receiving inadequate coverage and vital information would be missed. When planning my own fieldwork enquiries, it has been important to cover the whole area. For example, I did a beach survey where I collected pebbles and I decided to have my sampling points every 50 metres along the beach.

e This is a good answer and is certainly worth 5 marks against the maximum of 6 that could be awarded. Reference has been made to the method and the fact that it produces equidistant points. It also states the merits of the systematic method — it ensures equal covering, thus avoiding clustering. All of these points would be in the mark scheme for this question. When you attempt this paper, the examiners will assume that you have carried out your own geographical investigations and will credit references you make to your own experience when considering strategies and their relative merits. If this answer had indicated why sampling points were selected every 50 metres along the beach, then the full 6 marks would have been awarded.

(3) The business rate is a measure of the value of property and therefore, indirectly, of the land. Land values within a CBD are not easy to obtain directly, so this is a good way of obtaining that information.

e Again, this is a very good answer, particularly where only 2 marks are involved. The candidate could have written a more detailed explanation of what is understood by the term 'business rate'. Reference to the candidate's own fieldwork would also have been acceptable for credit.

(4)

Point number	Rateable value	Rank	Pedestrian count	Rank	d	d^2
35	4150	10	7	9	1	1
20	55000	4	10	8	−4	16
19	37700	5	242	2	3	9
18	172000	1	306	1	0	0
17	63000	3	225	3	0	0
16	20850	7	152	4	3	9
15	30300	6	112	5	1	1
12	90000	2	23	7	−5	25
10	18850	8	28	6	2	4
6	6150	9	2	10	−1	1

Σd^2 66

$$R_S = 1 - \left(\frac{6\Sigma d^2}{n^3 - n}\right)$$

$$R_S = 1 - \left(\frac{6 \times 66}{10^3 - 10}\right)$$

$$R_S = 1 - \left(\frac{396}{1000 - 10}\right)$$

$$R_S = 1 - \left(\frac{396}{990}\right) \quad 1 - 0.4 = 0.6$$

e Full marks are awarded. There are 4 marks available here which would be for:
- appropriate ranking (1 mark)
- correct calculation of the d^2 column (1 mark)
- correct calculations, using the formula with working in the spaces provided (2 marks)

(5) The Spearman rank coefficient of 0.6 is less than the critical value at the 0.01 level, but is greater than that at the 0.05 level. This means that as it is higher than the 0.05 significance level, the correlation between the two variables can be accepted, i.e. the higher the rateable value, the higher the pedestrian count.

e This answer receives all the marks available here. The only other information that could be given concerns the interpretation of the significance levels, i.e. there is a greater than 1% possibility but less than 5% possibility of the relationship occurring by chance.

Conclusions

It is not possible to provide a sample answer here as all the material that would be supplied in such a question is not given. We can, however, provide the following mark scheme:

Level 1: Simple statements made with reference to objectives or overall aim. Focused on a limited range of aspects. A poorly structured answer that has no clear pattern. No reference to own fieldwork in a way that relates to the title of this enquiry. Little idea of the limitations of the enquiry. No summary of findings or ways in which the enquiry could be extended. (1–3 marks)

Level 2: Statements must be developed. All objectives referred to in varying ways and linked to the aim/title in a clear order. Reference made to evidence. Awareness of the reliability of the information and ability to link this to own experience in conducting an enquiry. Reference to the candidate's own fieldwork must be meaningful to this enquiry and not just descriptive of the former. (4–7 marks)

Level 3: Development of points made in Level 2, but referring precisely and specifically to data collected as evidence. Limitations in all areas clearly stated. Awareness of the extent to which the aims have been realised. Some critical evaluation of the enquiry with meaningful suggestions as to its extension. Clear application of the candidate's own experience of fieldwork to areas such as limitations, the extent to which the aims have been fulfilled, and the extension of the work. As with all levels, fieldwork must be applied in the relevant manner to *this* enquiry. The answer must not be marked at this level if there is no reference to the candidate's own fieldwork. (8–10 marks)

Enquiry-related issues

This question relates to the development of enquiry-based fieldwork. This is covered under AQA (B) Unit 6.

Fieldwork-based questions

Each of the following questions (from a variety of examination boards) is asking you to write a report about a piece of fieldwork that you have completed. The precise details will therefore vary according to the enquiry that you have undertaken, and will reflect differing aspects of the material in the first part of this book.

Each question is followed by a brief synopsis of the mark schemes by which your answer will be assessed. There then follows an outline of a grade-A piece of fieldwork which could be applied to any of these questions.

AQA (B) Unit 6 — example question 1

Answer the following questions in relation to a geographical enquiry that you have carried out involving the collection of primary data in the field.

State the aim(s) of your enquiry.

(a) State two hypotheses or research questions which you established as part of the enquiry and explain why you chose them. (4 marks)

(b) Describe the methods you used to collect the data required to test the two hypotheses/research questions of your enquiry. (6 marks)

(c) Analyse the limitations of the methods described in part (b), and discuss the precautions that you took to ensure that the data were as accurate as possible. (7 marks)

(d) Discuss the extent to which the results of your enquiry supported your hypotheses/research questions. (8 marks)

e (a) A Level 1 response to this question would simply state the two hypotheses/ research questions and/or the concepts or ideas behind them. To access Level 2, more contextual detail would have to be offered.

(b) A brief description of methods would access Level 1 here, but to gain more credit, some detail of the methodology clearly linked to the hypotheses and location would have to be provided. Level 2 credit is therefore for the clarity of linkage(s) between data needed, method of collection, original hypothesis and field location.

(c) Here, limitations must be explained in detail to achieve the higher marks, and it must be clear how they applied to the situation described by the candidate. Precautions to ensure accuracy could relate to methods used to ensure accurate instrument readings, or the need for the averaging of several readings. Accuracy could be improved by better design of data collection methods, say of sampling techniques or questionnaire design. You may now begin to appreciate how the precise assessment of the answer depends very much on the quality of material offered by the candidate.

(d) This marks the logical conclusion of the enquiry. The question requires some appreciation of the geographical significance of the results of the enquiry, and of how they add to the understanding of the environment studied. References to results have to be provided, together with statements of conclusions that have been developed. References to anomalous results should also feature here, together with some evaluation of the overall success or otherwise of the enquiry. The best candidates are brutally honest here, and do not try to bluff the examiner. They give accounts of accurate results and findings, and state when and why the study went awry. It is very rare for a geographical enquiry to go completely to plan.

OCR (A) Unit 2682 — example question 1

Study the OS 1:50000 map extract provided.

Suggest a geographical investigation that you could carry out in an area shown on the map.

Your answer should include the following:

(a) A title for your investigation (either a question or a hypothesis) and location details. (10 marks)

(b) A description of the data to be collected and how you would organise their collection (include reference to any sampling scheme). (10 marks)

(c) A description of suitable methods of data presentation. (10 marks)

(d) A description of suitable methods of analysis. (10 marks)

(e) A description of secondary sources that would support the investigation. (10 marks)

e (a) To gain the highest credit, the title suggested must be both realistic for the area shown on the map and clearly linked to geographical theory. As stated in the question, the title has to be given in the form of a question or hypothesis.

(b) The data collection method must relate to the question given in (a) and be totally practicable. Clear discussion of the chosen sampling method should be given with consideration of time and/or spatial issues associated with that method.

(c) The chosen methods of data presentation must link logically to the data collected, and be appropriate for those data. As elsewhere, the depth of detail provided in the response will influence the range of credit awarded.

(d) As above, the chosen methods of analysis should be suitable for the data collected. An effective means by which conclusions can be generated from the analysis should be clarified. The method of analysis should follow on logically from the data collection and the original hypothesis.

(e) Simple statements of secondary sources would gain minimum credit here, but further commentary on how these sources would support the original investigation would take the answer to a higher level.

OCR (A) Unit 2686 — example question 1

You should use your report of your completed personal enquiry to answer the following questions.

(a) Geographical research commonly follows a series of steps like those outlined below:
  ▶ hypothesis or idea
  ▶ experimental design (sampling, measurement)

▶ data collection
▶ results and analysis (statistical test etc.)
▶ reconsideration of experimental design
▶ theory and explanation

Discuss the extent to which you followed similar steps in your enquiry. Where you deviated from them, explain why. (30 marks)

(b) How did you ensure that the data you collected were the most appropriate to the ideas/hypotheses you were testing? (30 marks)

(c) Describe and justify how you defined the geographical limits of your study area. (30 marks)

ℯ (a) Although there are six steps outlined in this question, your answer should contain three basic areas. Firstly, you should link the steps given in the question to your enquiry in a clear way. There should be a discussion of your enquiry with exemplification throughout. In other words, the context of your enquiry must be straightforward to see and relate to the steps given. Secondly, the steps given in the question must be followed as a complete research project. In particular, to achieve the highest grade, the fifth step, which recognises the need for review and evaluation, must be clearly and demonstrably understood. Finally, valid explanations for any deviations from the steps to the enquiry must be given, together with an explanation that a more full and detailed application of the steps may have improved the study overall.

(b) To achieve the higher levels of credit in this question, you should describe at least two ways in which you collected data, and demonstrate that each of these was appropriate to your study. You should also show a clear awareness of the limitations of your data collection process, as well as describing practical problems that may have been faced. To achieve the highest credit, evaluative commentary on the inherent problems of fieldwork design and experimentation should be given. In short, you must write about your data collection methods, but at all times recognise that problems of accuracy exist, and that improvements could be made under ideal, but often unrealistic, circumstances.

(c) All geographical studies have to be conducted within boundaries, which have to be both appropriate and practical. Here you should demonstrate that you selected an appropriate study area, and justify why the chosen area was right for the methodology of your enquiry. It would be wise to explain the processes that led to the selection of the area of data collection, preferably with reference to more than one set of criteria. For example, such criteria could include the ease of access to the area in question, the need for some uniformity of sampling population (say an area of common rock type) or, conversely, a need to sample a wider variety of data sources (for example, questionnaires in a number of different settlements). Higher marks would be gained if you could demonstrate an understanding that scientific enquiry demands careful control of the study area —

it should not be too small or too large, and should have clear and relevant parameters.

Edexcel (A) Unit 3 (Section B) — example

With reference to a piece of fieldwork that you have carried out on a physical geography topic:

(i) outline the aims of your fieldwork enquiry (3 marks)

(ii) draw an annotated sketch map of your study area to show where and why your data were collected (8 marks)

(iii) explain the sampling procedure that you used (3 marks)

(iv) explain the analytical methods you used to meet the aims of your investigation (6 marks)

> **e** At this point, this section is in danger of repeating itself. Hence, for
> (i) see the notes for AQA (B) Unit 6 (a)
> (ii) draw an accurate representation of the study area you used, and then refer to the notes for OCR (A) Unit 2686 (c)
> (iii) see the notes for OCR (A) Unit 2682 (b)
> (iv) see the notes for OCR (A) Unit 2682 (d)

Outline of a grade-A enquiry

Title
Is the climate in central Birmingham different from that of the rural–urban fringe?

Aim
The aim is to test the theory of urban heat islands by investigating the variations in temperature, humidity and wind speed between central Birmingham and the outer suburbs of the city.

Objectives
▶ To design a number of routes from the centre of the city to the urban fringe, identifying a number of sample sites on each route.
▶ To measure the climatic variables of temperature, humidity and wind speed at each of these sites.
▶ To prove the existence of the urban heat island and offer explanations for it.

Hypotheses
▶ Central Birmingham is warmer than the rural–urban fringe.
▶ Central Birmingham has a lower average wind speed than the rural–urban fringe.

▶ Central Birmingham has a lower relative humidity than the rural–urban fringe.

e Skills that could be demonstrated include:
- the drawing of location maps
- sampling techniques
- primary data collection — temperature, humidity levels, wind speed
- a variety of graphs — scatter graphs, regression lines
- Spearman rank correlation and its significance
- calculations of mean and standard deviation

The following would make the enquiry complete:
- good presentation of all results
- analysis of results
- commentary on findings of results
- commentary on limitations of data collection
- an overall conclusion
- a bibliography
- an appendix

Practical questions

AQA (B) Unit 6 — example

Look at Tables 1 and 2.

The data presented in Table 1 were collected from secondary sources as part of an enquiry relating to population and resource issues in countries in various states of development.

(a) Use the data for GNP and energy consumption (Table 1) to draw a scatter graph on logarithmic graph paper, showing the relationship between these two indicators. Identify each country by its reference number and insert a best fit trend line. (5 marks)

(b) Calculate a Spearman rank correlation coefficient for GNP and infant mortality rate. Comment on the statistical significance of the relationship. Critical values for R_S are shown below in Table 2. (5 marks)

(c) Assess the extent to which data on longevity are useful indicators of the level of development in a country. (4 marks)

(d) State *two* indicators, other than those shown in Table 1, that you would use to assess the extent to which a country is overpopulated. (1 mark)

(e) Write a brief report (no more than 300 words) to summarise the findings for this part of the enquiry. In the report you must:
- **(i)** discuss the nature of the relationships between the indicators
- **(ii)** evaluate the extent to which these indicators are appropriate in assessing the level of development (10 marks)

Country	GNP per capita (US$)	Energy consumption per capita (kg oil equivalent)	Infant mortality rate	Longevity — life expectancy at birth (years)
1 Kuwait	13 680	9200	15.6	73
2 Norway	20 020	5263	8.4	76
3 Saudi Arabia	6170	1900	71	63
4 Belgium	14 550	6049	9.2	74
5 Canada	16 760	9950	7.3	77
6 Japan	21 040	3680	4.8	79
7 UK	12 600	6369	9.5	75
8 Argentina	2640	1804	32	71
9 Mexico	1820	1227	50	66
10 Colombia	1240	685	46	66
11 Bolivia	570	367	110	53
12 Sri Lanka	420	106	22.5	70
13 Zaire	170	62	108	53
14 Malawi	160	56	130	49
15 Mali	230	27	117	45

Table 1 Economic and social indicators for selected countries (1990)

n	0.05 +/−	0.01 +/−
10	0.56	0.75
11	0.54	0.73
12	0.51	0.71
13	0.49	0.68
14	0.46	0.65
15	0.45	0.63
16	0.43	0.60
17	0.40	0.58
18	0.40	0.56

Table 2 Critical values for R_s at 0.05 and 0.01 significance levels

Sample answer for Unit 6 and examiner's comments
(a)

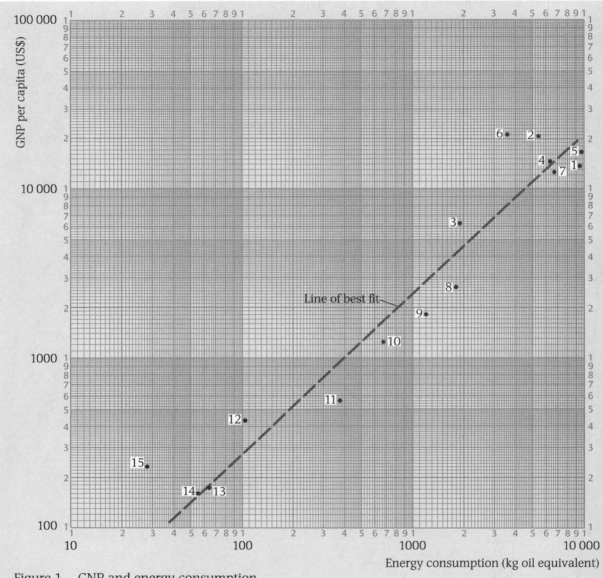

Figure 1   GNP and energy consumption

e As the graph paper provided is logarithmic on both axes, it does not matter which variable is placed on which axis. The 5 marks are awarded as follows:

- 1 mark for correct labelling of the axes
- 1 mark for correct use of the scales on these axes
- 2 marks for the plotting of the points
- 1 mark for the insertion of the best fit trend line

(b) The Spearman rank correlation coefficient is −0.90. This is significant at the 0.01 significance level. This means that the likelihood of this negative correlation occurring by chance is 1 in 100.

> **e** 3 marks are reserved for the calculation process and 2 marks for the commentary on the statistical significance.

(c) Longevity reflects the level of medical care in a country. The longer people live tends to be a measure of how rich or developed the country is because the country can afford to have good medical facilities. But longevity is not just how long somebody lives — it means that people are living longer over time in that country. Longevity increases as there are more medicines available, more hospitals and more opportunities for long-term care as people grow old. Longevity is also a response to the quality of diet in a country; the better the food people eat, the longer they will live. Longevity is therefore quite a good indicator of the level of development of a country.

> **e** This is a generalised answer which makes some valid points regarding medical care, diets and long-term care. The candidate could have referred to the data to identify possible anomalies to the overall trend. For example, people in Saudi Arabia have relatively low life expectancies, whereas those in Sri Lanka have longer ones. Clearly, other factors come into play, for example wealth distribution, incidence of preventable diseases and lifestyles. This answer gains 3 of the 4 marks available.

(d) The number of people who are starving, or who cannot read or write.

> **e** Any two indicators that suggest there are too many people for the resources available are acceptable. Other examples include food intake, number of patients per doctor, number of school places per 1000 people and so on.

(e) (i) The relationships between the indicators in the early part of this question demonstrate clear links. The relationship between GNP and energy consumption is positive and strong. There are some anomalies, such as Japan, where there is a lower use of energy than would be expected, and Kuwait, where the GNP is bigger than the trend. This is due to Kuwait being an oil exporter. The relationship between GNP and infant mortality is negative, but equally strong at −0.9, and this is highly significant. Saudi Arabia is an anomaly here as its infant mortality rate is nine times greater than that of Norway, yet its GNP is only one-third of that country. It also has a greater IMR than Mexico, even though it is substantially richer than that country. A relationship between longevity and GNP has not been attempted, and I do not have the time to do it, but it looks as though the richer countries have the longer longevities, which is what you would expect.

(ii) All of these indicators are appropriate for assessing the level of development. GNP is obviously used by many agencies to measure development across the world. Energy consumption indicates the amount of power used by the country, in industry, in houses and in transport, and so is an indication of wealth. However, colder countries will use more power than warmer countries. Infant mortality gives an indication of the health care provision in a country, and the amount of national wealth given to such things. Thus it indicates wealth and social considerations in a country. Longevity also reflects medical care, as well as the amount of medicines and quality of diet in a country. A high level of economic development is needed to pay for these.

290 words

e This answer gains maximum credit. In (i) the candidate has discussed the nature of each of the relationships between the indicators by both stating what the relationship is and pointing out anomalies to those relationships. The last relationship is described briefly, but does not detract from the quality of discussion earlier in the answer. In (ii) the candidate makes the more obvious points, but also qualifies the answer by pointing out possible reservations in using these indicators. For example, the answer states that temperature will influence energy consumption, and that health care provision is not just related to wealth, but also to perceptions of social welfare in a country. This sophistication of argument is what takes this answer to the highest level of response. Finally, note that the word count is given by the candidate. Had it gone over the suggested limit, marks would not have been deducted. The word limit is only a guide — examiners believe that longer answers are self-penalising, so any additional penalty is not necessary.

Edexcel (A) Unit 3 — example unit

In Unit 3 of this specification, you can choose either to submit a personal enquiry based on fieldwork or to enter for an examination based on fieldwork in which you are expected to 'demonstrate critical understanding and skills in an unfamiliar context'. For this, you are expected to undertake fieldwork exercises in both physical and human geography, as represented in the specification, and in the examination apply this personal experience to an unfamiliar situation and demonstrate your understanding of contrasting fieldwork investigations.

In this book, we have attempted to show you how such an examination would be presented. It has not been possible to give you all of the material that you could expect in the examination and this has obvious consequences for the example questions that can be set in this publication. As with other units, we have produced a number of questions and answers

that are representative of the types that will be set (and based on the material that we have been able to give to you). The examination paper itself would be longer.

Section A

Investigation aim

To analyse the impact of tourism in the area of Seaport

Investigation title

To what extent does the impact of tourism decrease with distance from Seaport town centre?

> In this exercise, you will be told that, in the context of this investigation, students obtained information from ten different sites along a coastal path. Information was obtained with regard to the numbers using the path and the amount of vegetation at each site.
>
> In the examination you would be given:
> - an Ordnance Survey map of the area
> - a page from a field notebook as recorded at one site by a student team
> - information on car parking at certain times during the day at selected sites, beginning with the town centre of Seaport (in bar graph format)

Example questions

	Site 1	Site 2	Site 3	Site 4	Site 5	Site 6	Site 7	Site 8	Site 9	Site 10
Number of persons per day	6	400	38	1200	300	57	212	19	52	3
Vegetation cover % per m^2	90	5	72	2	9	54	5	76	43	8

Table 1 People per day walking on the coast path near Seaport

(1) Using the information in Table 1, complete the graph in Figure 1 to show the percentage vegetation cover along the coastal path near Seaport.

(2 marks)

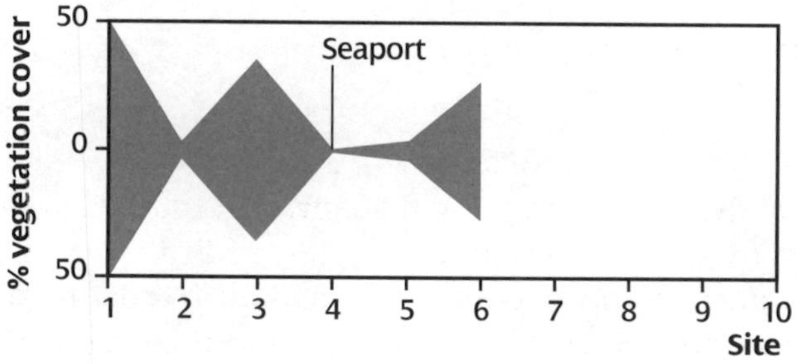

*Figure 1
Vegetation cover along the coastal path at Seaport*

(2) (a) Complete Table 2 below and calculate the Spearman rank correlation coefficient to show the relationship between vegetation cover along the coastal path and the number of people who use it.   (5 marks)

Site	Number of persons per day	Rank	Vegetation cover (% per m²)	Rank	d	d²
1	6		90			
2	400		5			
3	38		72			
4	1200		2			
5	300		9			
6						
7	212		5			
8	19		76			
9	52		43			
10	3		8			
						Σd^2

Table 2

$$R_s = 1 - \frac{6\Sigma d^2}{n^3 - n} =$$

Number of pairs of data	Significant levels 0.05	0.01
10	± 0.564	± 0.746
12	± 0.506	± 0.712

Table of significance values

(b) Assess the significance of the relationship by ticking a maximum of three of the following:

Negative	[ ]
Perfect	[ ]
Significant at 0.05	[ ]
Not significant	[ ]
Imperfect	[ ]
Positive	[ ]
Significant at 0.01	[ ]

(3 marks)

(3) In the context of the aim of analysing the impact of tourism in the area around Seaport:

(a) what other data collection would be appropriate in order to meet the aim of the investigation?   (3 marks)

(b) how could those data be collected? (3 marks)

(c) how could the collected data be represented? (3 marks)

(4) Using all the evidence that has been provided (including the OS map extract), comment on the hypothesis that 'the impact of tourism declines with distance from Seaport'. (15 marks)

Sample answers

Section A

(1)

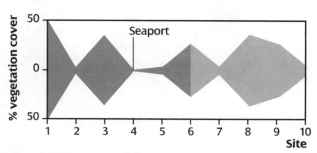

e The completion of such a diagram requires accuracy. If only 2 marks are available, the examiner may decide to remove marks from the maximum for every mistake that you make or, as is the case here, give $\frac{1}{2}$ marks for each site. The correct diagram must have:

- site 7 with 2.5% each side
- site 8 with 38% each side
- site 9 with 21.5% each side
- site 10 with 4% each side

(2) (a)

Site	Number of persons per day	Rank	Vegetation cover (% per m²)	Rank	d	d²
1	6	9	90	1	8	64
2	400	2	5	8.5	−6.5	42.25
3	38	7	72	3	4	16
4	1200	1	2	10	−9	81
5	300	3	9	6	−3	9
6	57	5	54	4	1	1
7	212	4	5	8.5	−4.5	20.25
8	19	8	76	2	6	36
9	52	6	43	5	1	1
10	3	10	8	7	3	9
					Σd^2	279.5

$$R_s = 1 - \frac{6\Sigma d^2}{n^3 - n}$$

$$= 1 - \frac{6 \times 279.5}{1000 - 10}$$

$$= 1 - \frac{1677}{990}$$

$$= 1 - 1.694$$

$$= -0.694$$

e The 5 marks here are awarded for:
- accurate ranking for numbers of persons and vegetation cover
- correct differences identified in the d column (including minus signs)
- correct calculations for the d^2 column
- correct calculation of the sum of column d^2
- application of the Spearman rank formula

(b) Negative   [✓]
    Perfect   [ ]
    Significant at 0.05   [✓]
    Not significant   [ ]
    Imperfect   [✓]
    Positive   [ ]
    Significant at 0.01   [ ]

e This is the correct answer, for 3 marks. The examiners are looking for:
- an understanding that the relationship is negative
- statement that the figure of −0.694 does not show a perfect relationship
- statement that the result is significant at the 0.05 level, i.e. the result shows that there is <5% possibility that the relationship established has come about by chance

(3) (a) To look at the impact of tourism in the area around Seaport, I would look at how beaches in the area have been affected. I would see how the quality of the beaches had been influenced by tourists and what sort of facilities had been provided and at what level.

(b) The information would be collected by surveying a number of beaches and using some sort of assessment based upon my observations. I would establish a number of categories, such as litter, and give each beach a score within those categories.

(c) Once the data had been collected, I would take maps of the area and draw bars to show the score for each beach. I would do this for each category. It would also be possible to produce a map showing a bar for each beach, giving the total scores that had been worked out in the survey.

e This answer would certainly, at AS, be awarded at least 2 marks for each section. To improve the answer in the first or second sections, the candidate could

have written more about the types of data that were going to be collected. Apart from litter, which was mentioned, other categories could include pollution (such as oil and tar), dog numbers and mess, visitor density, wildlife in the immediate area, ease of access and car parking provision. The candidate writes that an assessment would be carried out and 'I would give each beach a score' in all categories. A better answer would indicate how that score would be arranged, for example a point score running from −3 to +3, and also, as already mentioned, give some indication of the categories. The material given in the final section is good and would probably be credited with the maximum 3 marks. The candidate could have stressed, however, that the bars would be located at each beach surveyed, which would give an immediate spatial dimension to the survey once all of the beaches had been included on the map.

(4) —

e This type of summary question cannot be answered without all of the material which the examiners would provide.

Edexcel (B)

Edexcel (B) does not contain a coursework skills paper as such, although Unit 6 at A2 does give candidates 'opportunities to use a wide range of skills, including decision-making skills'. You are required, though, as part of the AS assessment in this specification, to submit a piece of coursework of not more than 2500 words based on an environmental investigation.

OCR (A) Unit 2682, Question 2 — example

(a) Study Figure 1, which shows a scatter plot of the number of pedestrians passing a point in a town during each of the time periods shown.

```
07.00–07.59    * *
08.00–08.59    * * * *
09.00–09.59    * * * * * * * *
10.00–10.59    * * * * * * * * * * * * * * * *
11.00–11.59    * * * * * * * * * * *
12.00–12.59    * * * * * * * *
13.00–13.59    * * * * * *
14.00–14.59    * * *
15.00–15.59    * * *
16.00–16.59    * * *
17.00–17.59    * * * *
18.00–18.59    * *
19.00–19.59    *
```
Each * represents 100 people.

Figure 1

Suggest the statistical measures that would be most appropriate to describe the distribution shown and explain how you would use them.

(20 marks)

(b) Table 1 shows values of discharge (cumecs) at different points along a stream. Describe and justify a method you might use to investigate the relationship between discharge and distance.

(10 marks)

Discharge (cumecs)	Distance downstream (km)
0	0
0.2	1
1.0	2
1.4	3
2.0	4
2.6	5

Table 1

Sample answers

(a) Figure 1 shows the number of pedestrians passing a point in a town, possibly a street corner, during a number of hourly periods. The purpose of this exercise may have been to see when there are more pedestrians in the town centre, but it could also have been used to see how pedestrian densities vary across the CBD of a town. This would be similar to an exercise I did on a day's fieldwork in Southport with my school, which was used to calculate where the edge of the CBD of Southport actually was.

In order to describe the distribution shown in Figure 1, I would use a variety of statistical techniques. These include measures of central tendency (mean, mode and median) and measures of variability or dispersion, such as the standard deviation and inter-quartile range.

The mean is difficult to calculate, as actual times are not given. The data are grouped, with one star equivalent to 100 people. This means that 6900 people walked past the point in the day. The mode can be seen more easily. The modal time period was between 10 a.m. and 11 a.m., which would again suggest that this was part of a shopping survey. The mode is identified as the biggest horizontal collection of stars. The median is the mid-point in the data, which means where the 35th star would be located when in rank order. The median is in fact between 11 a.m. and 12 noon.

Since you cannot calculate the mean accurately, the standard deviation cannot be calculated. However, looking at the data, there is a very clear positive skew, which means that the data are concentrated towards the lower times, in the morning. This can be confirmed by looking at the inter-quartile range. The lower quartile is at $(69 + 1) \pm 4 =$

17.5 (so it lies between the 17th and 18th star), and the upper quartile is between the 52nd and 53rd star.

The lower quartile is therefore between 10 a.m. and 11 a.m. (same as the mode, and only 1 hour less than the median), and the upper quartile is at 1300 hours to 1400 hours, some 2 hours after the median. This again confirms that the data are very much concentrated in the morning.

e This is a satisfactory attempt at this question. The candidate has concentrated on trying to use mathematical measures, and has realised how difficult it is to use the common descriptors of mean and standard deviation. An alternative approach would have been to consider graphical techniques to describe the data set, for example using a cumulative frequency curve and/or histogram. As the data are presented in a format similar to histograms, the mathematical techniques referred to are perhaps more valid. Another good point about this response is that the candidate has attempted successfully to use the measures identified to describe the distribution, and has even explained the meaning of the outcome.

The main area of weakness in this response lies in the opening paragraph. The candidate has not only repeated the context of the question, as given in the preamble, but has also related it to a piece of fieldwork he/she was involved in. This is not relevant to the question set and although not incorrect, does demonstrate inappropriate use of time. The time wasted in this manner might have been needed later in the examination.

As a range of statistical measures are identified, applied and commented upon, with some explanation of choice also being given, this answer receives a Level 3 mark of no less than 15.

(b) It can be seen from Table 1 that as distance downstream increases, then so does discharge. One way in which you could investigate this relationship is to use a correlation exercise, such as Spearman rank correlation. To do this, you have to rank the two sets of data in order from highest to lowest. The distance downstream in this case would be simply 5, 4, 3, 2 and 1, as would the discharge. You then have to calculate the differences between the two ranks, square them and put them into a formula. The answer then lies between 0 and 1, depending on how strong the relationship is. In this case, the answer would be 1, and the two sets of data would have similar ranks throughout.

Another method could involve drawing a scatter graph, which might have worked better in this case.

e This candidate has answered the question by correctly identifying an appropriate method to investigate the relationship between the two variables, discharge and distance. The answer then begins to explain how the technique works, and to apply it to the data provided. However, the candidate realises that in this case, the chosen technique would provide a very simplistic process — the answer would be 1, meaning total correlation. He/she then appears to lose confidence in the response, with a brief account of how the technique operates. The final sentence appears to confirm this opinion.

The candidate should realise that the question asks for an account of the chosen method, and not how it can be utilised in this case. He/she should therefore have given a thorough and detailed account of the method, and not worried about its application in this case. Finally, the answer would have reached the highest level by extending into an explanation of the meaning or significance of the outcome of that method.

In this case, the candidate has reached Level 2, scoring between 4 and 7 marks.

OCR (A) Unit 2686, Section B — example

(1) You have decided to make a study of industrial change over the last 50 years in the area covered by the settlements identified in Table 1.

	% employed in manufacturing	Total population (000s)
Holmfirth	40.5	21.1
Huddersfield	41.2	147.8
Brighouse	42.2	32.6
Cleckheaton/Liversedge	43.7	26.3
Mirfield	33.6	18.6
Dewsbury	38.1	49.6
Heckmondwike	44.3	9.8
Batley	41.4	45.5
Morley	32.0	44.6
Ossett	34.7	20.4
Horbury	29.4	9.8
Wakefield	25.2	74.8
Lofthouse/Stanley	28.1	17.4
Leeds	27.2	445.2

Table 1 Urban areas in parts of West Yorkshire

(a) Classify the urban areas according to the percentage employed in manufacturing and construct a suitable key for a choropleth map. Justify your choice of number of classes to use and shadings for them on the map. You are not to produce the map. (10 marks)

(b) Describe how you would construct a dot map for the population in each of the urban areas listed in Table 1. State what dot value you would choose and justify your choice. You are not to produce the map. (10 marks)

(c) A field survey of a sample of buildings used for manufacturing in some of the towns in the area produced the results shown in Table 2.

(i) State a hypothesis to investigate the changes in use of manufacturing buildings of different ages.

(ii) Name a statistical test that could help demonstrate a difference between the use of buildings of different ages and describe how you would apply it. You are not to do the actual calculation.

(10 marks)

	Remaining in original manufacturing use	Manufacturing type change, or not used for manufacturing
Pre-1945 buildings	6	31
Post-1945–1985 buildings	19	12
Post-1985 buildings	24	2

Table 2 **The use of buildings originally erected for manufacturing**

(2) You are going to carry out a visitor survey at a popular recreational site in the Midlands to find out where tourists come from and why. You have obtained the data from a previous survey on the origin of visitors shown in Table 3.

(a) Suggest and justify a method of representing the data in Table 3 on a map of the UK. (5 marks)

(b) What factors might help explain the pattern of visitor origins?

(10 marks)

(c) You have been given permission to survey visitors in the car parks. Justify five questions you would ask and the steps you would take to ensure that you obtain a representative sample of visitors. (15 marks)

UK region	% total visitors	UK region	% total visitors
Midlands	52	Southwest	2
Southeast	15	Wales	2
Yorkshire/Humberside	12	Cumbria/Northwest	1
Lancashire	8	Scotland	1
East Anglia	3	Northern Ireland	1
Northeast	3		

Table 3 **Origin of UK visitors, by region, to a Midlands Heritage site**

Sample answer

(1) (a) There are 14 towns listed in Table 1. In order to classify them into groups, for a choropleth map to show the percentage employed in manufacturing, I would look at the bottom percentage given in the table (25.2) and the top percentage (44.3). This means that there is about 20% between these upper and lower values. As it is always better to have about four or five classes, I would subdivide these numbers by 5% at a time. So my classes, and the number of urban areas in each class, would be:

25.0–29.9	4
30.0–34.9	3
35.0–39.9	1
40.0–44.9	6

Note that I have used .9 in my class boundaries, so as to ensure that all possible elements within the data set are included, and also to avoid any overlap between classes. It would have been wrong, for example, to have 25–30 and 30–35, as there would have been overlap at the number 30.

Using these classes, I would then draw a choropleth on an outline map, using the following key:

25.0–29.9	
30.0–34.9	
35.0–39.9	
40.0–44.9	

I have chosen this key so as to use a darker density of shading for a higher number, and a lighter shading for a lower number. I would not use black as this would make it difficult to write over, and I have not used white as this is often used to indicate no data.

e This question has asked the candidate to complete four tasks for 10 marks. The four tasks are to classify the data, construct a key for a choropleth map, and to justify both of these. Therefore, it is safe to assume that there are at least 2 marks for each of those tasks, with an additional 2 marks for further development.

This candidate not only classifies the data, but also applies them and presents a frequency table. The system by which the data are classified is not scientific, but appropriate. The usual way by which the number of classes are determined is by the formula:

number of classes = 5 × the logarithm of the total number of items in the data set

= 5 × log14 = 5.7 (rounded up to 6)

However, in this case, the number of classes chosen (4) is acceptable. The candidate even goes on to justify why use of the decimal point is important in this case. The data are presented in continuous form, and hence the class boundaries should reflect this. There is no potential for overlap or for gaps. The shadings given are also suitably graduated, with a darker system being used for higher percentages, and there is also good justification for the system provided. This candidate satisfies all four tasks, giving further development on reasons for choice, and would gain full marks.

(b) A dot map is constructed by allocating one small dot for a fixed number of the item you are trying to map, and then placing those dots as evenly as possible in the area given on the map. In this case, the data are population, so one dot would correspond to so many thousand people. The largest city is Leeds (445 000) and the smallest is Heckmondwike and Horbury (nearly 10 000). I would therefore say that one dot should equal 10 000 people, rounded up or down to the nearest 10 000. Therefore, Leeds would have 44 or 45 dots and the second largest town, Huddersfield, 15 dots. I have chosen this dot value as it is easy to calculate, and does not provide too many dots to place in any one area. Also, the smallest two towns would have a dot too.

e As in (a), there is a variety of tasks to be undertaken: the explanation of map construction, the identification of dot value and the justification for this. The candidate clearly understands how to construct a dot map, but more detail of methodology could have been offered.

The candidate makes the choice of dot value in the context of the range of population totals given in the table, and also refers to the use of the space available, even though that is not provided. The dot value offered (one dot to 10 000 people) is perhaps at the upper end of the scale, and may have been better at one dot per 5000. The rationale, namely that it is easy to calculate, falls down a little when applied to Leeds. The candidate states 'Leeds would have 44 or 45 dots' — which is it? Accurate reference to the data would confirm that, according to this candidate's classification, Leeds would have 45 dots. Finally, it is best to represent even the smallest of areas with at least two dots — how do you place just one dot evenly in an area? This would also mean an ideal dot value of 5000.

Overall, this is still a good answer, scoring 8 of the 10 marks available.

(c) (i) It would appear that as the buildings get older, they are less likely to be still in use for manufacturing. A hypothesis could be: 'As buildings originally erected for manufacturing get older, they are more likely to be used for activities that do not include manufacturing. They also may have changed their manufacturing type'.

e The candidate has recognised that there has been a greater change of use in the older buildings than in the newer ones, and consequently will gain credit for the hypothesis stated. However, the candidate has not fully understood the final column of the data. The buildings may still be in manufacturing use, but that use may be different from its original manufacturing function. A second sentence to the hypothesis hints at this, but it occurs as an afterthought and spoils the first sentence. This confusion also continues into part (ii) — see below.

   (ii) One technique that could be used to demonstrate a difference between the use of buildings of different ages is the chi-squared test. This is used to test how different a set of data that has been observed is from one that would be expected to occur. In this case, of the 37 buildings older than 1945, 6 are still in use for manufacturing and 31 are not. It may be expected that half would have changed, so the expected numbers would be 18.5 for each. The same technique could then be used for the buildings from 1945–85 and post–1985.

   Having calculated the expected values, they can be compared with the observed values and the difference between them found (O − E). In the case of pre-1945 buildings, these values would be −12.5 and +12.5. Similar results would be identified for the other two time periods. These differences are then squared and put into a formula. The answer is then compared with statistical tables to see how significant the result is.

e The candidate opts to name the chi-squared test, which is both valid and appropriate. However, this is a very difficult test to comprehend fully, both in terms of its application and the interpretation of its result. Despite the instruction not to do the actual calculation, the candidate attempts to explain the technique by applying it to the data, but soon stops doing so, perhaps realising that he/she may not be applying it correctly. The answer then reverts to a general description of the technique, referring to a formula and statistical tables in very vague terms. The candidate appears to have lost confidence in his/her ability to describe how the technique works.

More detail could have been offered regarding the need for a table to show the differences between observed and expected values, the precise details of the formula could have been given, and more detail of the procedures for using the statistical tables could have been offered. Overall, this candidate scores 6 of the 10 marks available. The lesson is clear here: ensure that you understand fully how statistical techniques can be applied.

(2) (a) I would take a map of the UK, divided up into the different regions as given in the table, and I would draw a bar graph, to scale, coming up out of each area. So, for example, I would use a scale of 1 cm = 10%, so a bar that was 5.2 cm would rise up out of the Midlands, and 1.5 cm from the Southeast and so on. I would use this, as it is easy to construct, can be drawn to scale, and would give a clear indication of where most visitors come from to the Heritage site. As would be expected, most of the visitors come from the surrounding area.

e This candidate has given a clear indication of a very simple technique to represent the data and has explained how the technique would be applied. Despite the simplicity of the technique, it is highly appropriate to the task and is therefore perfectly valid.

More 'sophisticated' answers might have suggested a choropleth map, but closer examination of the data would highlight the problems caused by the wide range in numbers in the data. Class sizes would have to be uneven, and the smallest class, say 0–5, would have contained over half of the data. As a visual representation, a choropleth map may not be as effective as the method described by the candidate.

This candidate also justifies the choice of technique in straightforward terms and hence would gain 5 marks for this answer.

(b) There are a large number of factors that might explain the pattern of visitor origins. The most important factor would be how accessible the site is for visitors. If the recreational site is on a main road, or near to a motorway or railway, there will be more visitors to that site. People can travel to the site easily and without too much hassle, and so more people will be able to come. This site seems to attract a lot of people from nearby in the Midlands, although there are a few people from long distances away, such as Northern Ireland and Scotland.

Another factor would be what type of recreational site the place is. If, for example, it is a theme park like Alton Towers, people will be prepared to travel large distances to the site, as they know they will have a good time there, and be able to spend a long time there. There are activities at Alton Towers to keep people of all ages occupied — for example, rides for the 'young at heart' and gardens for the older people. If the site is a historical site, then again people may be attracted from long distances, but only if they are interested in that topic. Others may be staying in the area to visit a number of such sites. They may be on holiday, for example.

A third factor would be price of entry to the site. If the site is relatively cheap, more people will visit it — there may be a family ticket reduction which would encourage more families to visit the site. If this were the case, it could result in more people being attracted from nearby.

A final factor would be how well advertised the site is across the country. If it has good publicity in newspapers, brochures or on TV, this will encourage more people to visit it, both from nearby and from further afield.

e This candidate has recognised that there are a number of factors that could be relevant. A minor point should be noted here: it is better not to state 'a *large* number of factors' in an introductory sentence when, in fact, you only identify four. It is better simply to state that there are 'a number of factors'.

Each of the four factors is valid, and the candidate has to a greater or lesser extent developed them in turn. The accessibility factor is quite well argued, but there could have been more precise references to relative travel times on roads versus motorways, the importance of motorway junctions and car parking. The very brief reference to railways is rather simplistic — a nearby station is needed for a railway line to be influential. The point relating to the nature of the site is also quite well developed, although more could have been made of the holiday versus day trip aspect. The third and fourth factors are also clearly, if simplistically, described.

The candidate does, however, provide an example, with the reference to Alton Towers, and this is well stated.

This response gains 8 or 9 marks out of 10.

(c) I would ask the following five questions:
- Please would you tell me where you have come from? I would like to know your town or the first part of your postcode.
- How long did it take you to get here?
- How long do you intend staying here? (Or, if after the visit, how long did you stay?)
- What is your age range? (I would offer a number of age ranges, such as under 16, 17–35 and so on for them to select from, so as not to upset them.)
- Why did you come to this site?

I would ask the first and second questions because a survey of visitor patterns would have to include these; otherwise it would be pointless. You need to know how far people have travelled, in terms of both distance and time. The third question gives an indication of how popular the site is, and how attracted the people are to the site. The fourth question gives some insight into the nature of the people coming to the site, to see if it is more attractive to a certain group of people. The first four questions are all closed questions, which would make it easy to process the answers. However, the last question is more open-ended, and allows the person to provide information as to why he/she came. This may relate to being part of a holiday or visiting friends and relatives, or could provide information on how well advertised the site is.

To obtain a representative sample of visitors, I would use systematic sampling. I would stop every tenth car visiting the site at the entrance to the car park or leaving it at the exit. I would carry out this sampling on a number of different days, so as to get a more even spread of visitors — sometimes weekdays, sometimes weekends. The question implies that there is more than one car park, so I would visit each one.

e The five questions suggested by this candidate are all acceptable. As the candidate recognises, there is a blend of closed and open-ended questions. Other data that could have been collected include the number of previous visits to the site, or to similar sites, the nature of the family grouping in each car, and which parts of the site were most enjoyable (if conducted after the visit). It is pleasing to see that use is *not* made of inappropriate questions, such the sex of the person, the income, or even the means of transport to the site (the questions are being asked in a car park). The justification for each question is also good and this part of the answer gains high marks.

The section dealing with the sampling method is weaker. The sampling method described is unlikely to work in practice — people will not want to be stopped either entering or leaving the car park in their cars. The questions would be best conducted in the car park as people move to and from their cars. Some stratification of the sample, by age group for example, would help to ensure that each visitor type is included. There are some good features, though. The candidate recognises the need to sample on a variety of days, and to go to different car parks, so as to avoid the impact of bias as much as possible.

Overall, this is a good answer, gaining 13 or 14 marks out of a possible 15.